# Exploring Urban Change in South Asia

The series incorporates work on urbanisation and urbanism in South Asia from diverse perspectives, including, but not being limited to, sociology, anthropology, geography, social policy, urban planning and management, economics, politics and culture studies. It publishes original, peer-reviewed work covering both macro issues such as larger urbanisation processes, and economic shifts and qualitative research work focused on micro studies (either comparative or ethnographic based). Both individual authored and edited books are considered within the series with the possibility of identifying emerging topics for handbooks.

Ram B. Bhagat · Mohammad Izhar Hassan

# Urbanisation and Urban Policies in India

## Trends, Patterns and Emerging Perspectives

Springer

Ram B. Bhagat
Department of Migration and Urban Studies
Formerly at International Institute for Population Sciences
Mumbai, Maharashtra, India

Mohammad Izhar Hassan
Department of Geography
Formerly at Maharshi Dayanand University
Rohtak, Haryana, India

ISSN 2367-0045 ISSN 2367-0053 (electronic)
Exploring Urban Change in South Asia
ISBN 978-981-96-3183-4 ISBN 978-981-96-3181-0 (eBook)
https://doi.org/10.1007/978-981-96-3181-0



This Springer imprint is published by the registered company Springer Nature Singapore Pte Ltd.
The registered company address is: 152 Beach Road, #21-01/04 Gateway East, Singapore 189721, Singapore

# Preface and Acknowledgements

We intend that this book serves as a text book for post-graduate students and researchers at the university level who are interested in the issues of urbanisation as linked to migration and several other spatial issues resulting from spatial transition. The book has been kept simple as far as possible. It is the result of our experiences of teaching over three decades in various universities and institutes such as Maharshi Daya Nand University, Rohtak; Ravenshaw University, Cuttack; International Institute for Population Sciences, Mumbai; Tata Institute of Social Sciences, Mumbai; Gokhale Institute of Politics and Economics, Pune, and lectures delivered in India and abroad.

Urbanisation is a multi-disciplinary subject, but very much theoretically founded in space and place approach. It is about changing relationships between people and place. While spaces are produced, places are made having history, culture, society, polity and economy. Urbanisation is about making of places. It is not a finite process but urbanisation may be followed by deurbanisation and reurbanisation historically. Therefore, making and unmaking of places are crucial challenges for urban policy and programmes for inclusive and sustainable urbanisation. Our effort has been to focus on demographic components and migration, census towns and *in situ urbanisation*, access to basic services by size class of urban centres, informal settlements, urbanisation and regional disparities, and urban policies and programmes as factors influencing spatial transition. Spatial transition is very crucial for who gets what and where, closely related to inclusive and sustainable urbanisation. The book highlights that urban policies can play a very important role in spatial and distributive justice. However, there is a need to make a distinction between urban development and urbanisation policies, and make appropriate strategies for them. We hope that the book will serve our objective and urban imagination of right to the city highlighting the fact that better life is possible only in better places.

We are conscious of the fact that the nature of urbanisation contributes to the changing nature of politics and sociology as well. We would like to mention that politics originated in *polis,* and the metropolises dominate economy, society and polity in our contemporary times. As such, social and economic inequalities manifest in regional and spatial inequalities shaped by the processes of urbanisation. We hope that our understanding and the interpretation of urban demographic change shall encourage research on urbanisation and urbanism in other disciplines as well.

In writing this book we owe a great deal to our colleagues, friends, students and well-wishers. We are very thankful to Prof R.N. Sharma, former Dean, School of Development Studies, Tata Institute of Social Sciences (TISS), Prof Ajit Ranade, former Vice-Chancellor, Gokhale Institute of Politics and Economics, Pune; Prof. Amitabh Kundu, former professor, JNU, New Delhi, Prof. Ravi Srivastava, former professor, JNU, New Delhi, and Prof. Alakh N. Sharma, Director, Institute of Human Development, New Delhi; Prof Ashwini Kumar, Dean, School of Development Studies, TISS; Prof Abdul Shaban, TISS, Mumbai, Prof S. Irudaya Rajan, former professor, Centre for Development Studies, Trivandrum; Prof Sachchidanand Sinha, former professor Jawaharlal Nehru University, Prof Sayeed Unisa former professor, International Institute for Population Sciences, Mumbai and Prof. Pritirekha Daspattanayak, former professor Ravenshaw University, Cuttack for their support and keen interest in our work. Our constant interactions with them shaped our thoughts, ideas, and consciousness over the years. Prof K.R. Dikshit, former Head, Department of Geography, University of Pune and patron, Institute of Indian Geographers has always been a source of inspiration and guidance for us. We are very thankful to him for his goodwill and good wishes. Our special thanks to Dr Praveen Kumar, Library and Information Officer, and Mr D.N. Vidhate, Assistant Library and Information Officer, Mr Rakesh Patil, Mr Santosh Pawar and Mr Ashish Patole, Library Staff IIPS Mumbai for their timely help and support.

Our thanks are also due to Ms Satvinder Kaur, Commissioning Editor and Ms Ramya Somasundaram, Project Coordinator, Springer Nature in preparing the manuscript in the required format. We would also like to place on record our sincere thanks to Ajay Murmu, Inakoti Prasad and Vasim Ahamad, PhD Scholars, IIPS for their help in collection of some relevant literature and for drawing maps. Last but not the least, we appreciate the support and encouragement from our family members and we express our profound admiration for their forbearance and understanding during writing this book.

Ram B. Bhagat
Mohammad Izhar Hassan

**Competing Interests** The authors have no competing interests to declare that are relevant to the content of this manuscript.

# Contents

# About the Authors

**Ram B. Bhagat** is former Professor and Head, Department of Migration and Urban Studies, International Institute for Population Sciences (IIPS), Mumbai. He has also served as Lecturer (Assistant Professor) and Reader (Associate Professor), Department of Geography, Maharshi Dayanand University, Rohtak, Haryana (India). Professor Bhagat has been teaching and doing research on population, migration and urbanisation, ethnicity and politics, citizenship and public policies over three decades. As a professional of migration and urban studies, he has participated and contributed in various international and national forums of governmental and non-governmental organisations including various UN bodies namely UNESCO, UNICEF, IOM, ILO, and Population Division, DESA/UN, apart from delivering lectures in India and abroad.

**Mohammad Izhar Hassan** is former Professor of Geography at Maharshi Dayanand University, Rohtak, Haryana (India). After obtaining Ph.D. from JNU, New Delhi, he began his academic career in the year 1991. His area of interest is population studies in geography. More than two dozen research scholars have obtained their Ph.D./M.Phil. degrees under his supervision. He has also served as Professor of Geography at Ravenshaw University, Cuttack (Odisha) during 2014–17.

# List of Figures

# List of Tables

# Chapter 1
# Nature of Urbanisation—Historical and Global Perspectives

## 1.1 Urbanisation: A Historical Perspective

Although the nature of urbanisation that we are acquainted with has its roots only in the recent past, i.e. post-Industrial Revolution in the middle of the eighteenth century, the earliest form of settlements based on distinct non-agricultural pursuits is said to have appeared more than five thousand years ago. Human settlements came into existence after the Neolithic revolution that occurred some 10 to 12 thousand years ago. Domestication of animals and plants was the turning point in human history that led to the introduction of sedentary farming and permanent human settlements. With time some of these human settlements underwent a significant transformation in their economic function in the wake of development in the process of cultivation practices. However, this transformation of rural settlements into urban settlements was not a sudden development nor permanent agriculture was the only reason underlying the emergence of cities or towns. Archaeological evidence indicates the emergence of the first urban centre in Mesopotamia sometime around 3200 BC, and there was a gap of about five thousand years between the appearance of permanent agriculture and the emergence of the first 'city' on the earth (Tellier, 2019: 20). Improvements in agricultural practices over time gradually led to surplus production of food including dairy products. A part of the agricultural labour force was diverted to non-agricultural pursuits. Thus, there appeared a system of exchange of agricultural produce with outputs having origin in non-agricultural activities. Initially, this system was highly localised, and the emergence of urban centres had to wait for several thousand years as the mobility of people as well as goods was highly restricted. However, improvements in the transportation system with the invention of wheels, wagons and boats, that took place during 4500–3000 BC in the Black Sea region of Western Asia, and that later spread to other areas, paved the way for long-distance trade and commerce (Tellier, 2019: 21). Invention of writing that is said to have occurred in ancient Sumer around 3600 BC (Murphey, 1970: 775) further revolutionised long-distance trade. The development of writing enabled record-keeping and complex accounting



R. B. Bhagat and M. I. Hassan, *Urbanisation and Urban Policies in India*, Exploring Urban Change in South Asia, https://doi.org/10.1007/978-981-96-3181-0_1

apart from facilitating the sending of long and elaborated messages to distant locations. Remarkably, these innovations and developments and also many others like the plough, wind-mill, potter's wheel, etc., took place mainly in this region of Western Asia, and it is from here that they gradually spread to other areas of the globe. These developments are said to have played a far more important role in the process of the emergence of urban 'cities' than only the production of surpluses in agriculture.

Development of the concept of 'state' and, later formation of 'city-states' were also instrumental in the emergence of urban system. In fact, urbanisation had several cradles, and the emergence of first 'urban cities' in Mesopotamia and a little later in Egypt and elsewhere around the Mediterranean in the fourth millennium BC, was followed by similar developments in the Indus Valley region and along the Yellow and Yangzi rivers in China roughly after around a millennium later. Urban life also developed in the coastal zones and low-lying valleys of the Andean region and Mesoamerica, and elsewhere in different parts of the world. Notwithstanding the existence of settlements with distinct urban character in different parts of the world in the ancient times, the transformation from rural to urban remained very slow for several thousand years, and until very recently in world history, almost the entire population lived in the rural areas. The few cities and towns that were in existence in different parts of the world appeared nothing more than tiny islands in the ocean of rurality (Hassan, 2020: 51). In fact, these urban centres were not only very small in size but also many of them faced decay and were later abandoned under different circumstances. Thus, the number of urban centres as well as the degree of urbanisation remained very low even in the most advanced regions of that time at least until 1000 BC. Persisting labour-intensive farming techniques, high death rates in the lack of medical advances, restricted rural–urban migration, absence of large-scale manufacturing, bureaucratic control of peasantry, widespread traditionalism and religiosity that retarded technological and scientific advancements in association with inadequate political control of nobility in the hinterlands were the underlying reasons for an extremely slow pace of rural–urban transformation (Davis, 1955: 431–32).

Towards the middle of the first millennium BC, the process of urban development gathered momentum in the Greco-Roman region of Europe lying to the north of the Mediterranean Sea. Development of iron tools and weapons, alphabetic writing, improved sailboats, cheap coinage, more democratic institutions and systematic colonisation that had significant bearings on production, trade and administrative and political control were the underlying reasons (Davis, 1955: 432) for this fresh wave of urbanisation. Towns and cities not only became more numerous but also grew in size leading to an increased share of the population living in the urban centres during the reigns of the Romans. Athens, Syracuse and Carthage were some of the important cities of that time. When at their peak, the Romans ruled over a vast territory surrounding the Mediterranean starting from north-western Europe, and covering parts of southern Europe, Western Asia and northern parts of Africa in a clockwise direction along the Mediterranean Sea. The Roman power came to an end towards the beginning of the fifth century AD. The region could boast of the existence of some of the largest cities of the world, e.g. Rome and Constantinople (the latter is presently known as Istanbul) till the rise of cities like London in the nineteenth century after the Industrial Revolution.

However, the true urban revolution with a substantial proportion of the population residing in the urban areas was not witnessed until the onset of the Industrial Revolution in north-western Europe in the middle of the eighteenth century. The urban centres that came into existence at different points in time, and in different parts of the world, prior to the Industrial Revolution were by and large tied to an economy that was primarily based on agriculture with handicrafts occupying only a secondary position (Davis, 1955: 432). The initial impetus came from revolutionary changes in farming practices that led to a massive rise in production and productivity. Subsequently, improvements in transport, discoveries of new lands and new trade routes and finally the introduction of factory-based and machinery-driven production of industrial goods gave a massive boost to urban transition in Western Europe. These changes gradually spread to the rest of Europe and other present-day developed countries outside Europe. The rate of growth in urban population became distinctly higher than that in rural areas. Migration from the countryside to towns and cities played a key role in increased urban transition. With improvements in food supply and development of medical sciences, mortality decline, in general, and in rural areas, in particular, triggered population growth which induced a shift of population from rural areas to towns and cities (Hassan, 2020: 51).

## 1.2 Global Urbanisation: Trends and Patterns

The technological explosion associated with the Industrial Revolution led to an unprecedented rise in the process of urbanisation. Manufacturing and commerce became increasingly important activities of urban life, and there was a massive migration of the population from countryside to industrial cities. North-western Europe, particularly England, was the hub of this new wave of urbanisation. London soon emerged as the leading city, and by the turn of the nineteenth century, the population of the city was already close to a million mark. Britain became the most urbanised country in the world, and the level of urbanisation that Britain had achieved by 1801 was not reached by any other country till 1850. Other European countries followed roughly after 1850. Around the same time urban transition also spread to other parts of the world outside Europe which are presently known as the developed world. In the rest of the world, urbanisation could pick up only towards the mid-twentieth century.

According to the available estimates till around 1800, only about 3 per cent of the world's population lived in towns and cities. At that time the world had less than 50 cities with a population of 100,000 or more (Davis, 1955: 434). By the middle of the nineteenth century, however, the share of the population residing in urban areas more than doubled and the level of urbanisation is reported to have reached 6.5 per cent in 1850. By the year 1900, nearly 16 per cent of the world's population was urban, and there were as many as 16 cities with a population of more than 1 million each

(Hassan, 2020: 51). The period of a hundred years from 1850 to 1950, particularly its later half, witnessed a decided shift in rural–urban balance in human history, and by 1950 as much as 29 per cent of the world's population was urban (Lowry, 1990: 150). By contrast, nearly 60 per cent of the population in the high-income industrialised countries of the world was already urban in 1950. Around this time there were nearly 900 cities with a population of 100,000 or more, and this number is larger than the number of urban centres with a population of even 5000 in the year 1800 (Davis, 1955: 434). Towards the close of the twentieth century, nearly 47 per cent of the population was already urban, and by 2007 world population was, for the first time in human history, evenly distributed between urban and rural areas (Hassan, 2020: 51). At present more people, i.e. 55 per cent population of the world, live in urban areas than in rural areas, and by the year 2050, 68 per cent is projected to be urban (UN, 2019).

## 1.3 Regional Pattern

There exists a marked variation in the urbanisation levels reached by different geographic regions and countries across the world. The most urbanised geographic region of the world is Northern America followed by Latin America and the Caribbean with more than eight-tenths of their population as urban (Table 1.1). Europe comes next with an urbanisation level of 74.5 per cent. The pace of urbanisation in Latin America and the Caribbean was exceptionally rapid during the latter half of the last century as a result of which the share of urban population increased from just a little over 40 per cent in 1950 to over 80 per cent in 2018 implying an average annual rate of change of over 1.4 per cent as against 0.6 per cent in Europe. With this remarkably rapid pace of urbanisation, Latin America and the Caribbean equalled the level of urbanisation of Europe in 1988, and thereafter pushed Europe to the third position. On the other extreme, while Asia is about to reach a balance between urban and rural populations only now, in Africa only a little over 43 per cent population is urban, up from 35 per cent in 2000.

There exists a positive association between levels of economic development and per cent population residing in urban areas in individual countries and also across regions within a particular country. The more developed parts of the world are more urbanised than the less developed areas although the gap in the levels of urbanisation of the two is narrowing down over time. At the aggregate level, while high-income countries were already 'urban-majority' prior to 1950, the middle-income countries achieved this distinction only in the year 2014. For the low-income countries the population is expected to be evenly distributed between urban and rural areas not until the middle of the present century.

Urbanisation in the less developed parts of the world gathered momentum only towards the middle of the last century, and by then most of the high-income developed countries were already 'urban majority'. The nature and process of urbanisation occurring now in less developed countries differ from what was earlier experienced

**Table 1.1** Mid-year population residing in urban areas—world and major regions

| *World/Major Regions* | *Per cent Population Residing in Urban Areas* | | | | | |
|---|---|---|---|---|---|---|
| | *1950* | *1975* | *2000* | *2018* | *2030* | *2050* |
| World | 29.6 | 37.7 | 46.7 | 55.3 | 60.4 | 68.4 |
| More developed regions | 54.8 | 68.8 | 74.2 | 78.7 | 81.4 | 86.6 |
| Less developed regions | 17.7 | 26.9 | 40.1 | 50.6 | 56.7 | 65.6 |
| Less developed regions, excluding China | 20.3 | 30.9 | 41.2 | 48.0 | 53.1 | 62.7 |
| High-income countries | 58.5 | 70.4 | 76.8 | 81.5 | 83.9 | 88.4 |
| Middle-income countries | 19.9 | 29.4 | 41.6 | 52.6 | 59.0 | 68.3 |
| Low-income countries | 9.3 | 17.5 | 25.7 | 32.2 | 38.3 | 50.2 |
| Africa | 14.3 | 24.7 | 35.0 | 42.5 | 48.4 | 58.9 |
| Asia | 17.5 | 25.0 | 37.5 | 49.9 | 56.7 | 66.2 |
| Europe | 51.7 | 65.5 | 71.1 | 74.5 | 77.5 | 83.7 |
| Latin America and the Caribbean | 41.3 | 61.0 | 75.5 | 80.7 | 83.6 | 87.8 |
| Northern America | 63.9 | 73.8 | 79.1 | 82.2 | 84.7 | 89.0 |
| Oceania | 62.5 | 70.9 | 68.3 | 68.2 | 68.9 | 72.1 |

Source *World Urbanization Prospects: The 2018 Revision*, United Nations, Department of Economic and Social Affairs, Population Division. Available at https://population.un.org/wup/
*Notes* 1. More developed regions comprise Europe, Northern America, Australia/New Zealand and Japan. 2. Less developed regions comprise all regions of Africa, Asia (except Japan), Latin America and the Caribbean plus Melanesia, Micronesia and Polynesia. 3. The country classification by income level is based on 2016 GNI per capita from the World Bank

in developed countries in many ways—first, the scale and pace are unprecedented; second, there is a mismatch between average income levels and the pace of urban transition in individual countries; and finally, nature and direction of urban transition are more linked with the global economy now than before (Cohen, 2004: 24). Currently, urban population in the less developed parts of the world is growing much faster than in the developed world leading to a decided shift in the balance of urban population between them. In fact, divergence in urban growth rates of the two worlds had set in during the first half of the last century itself. About 60 per cent of the world's urban population came from the developed regions in 1950, but by the close of 1960s the same was equally divided between the developed and the less developed region. In 2018, more than three-fourths of urban dwellers of the world lived in the less developed regions, and by 2050 the corresponding figure is projected to increase to 83 per cent. This implies an ever increasing share of the less developed regions in the annual increment of world urban population since 1950. While in 1950–55, the less developed countries accounted for 57 per cent of the annual increment, by 2015–20 their share was as high as 94 per cent (UN, 2019).

In the wake of this rapid growth in urban population, the world has seen not only a massive increase in the number of large cities but also the emergence of cities of unprecedented sizes. The average size of the world's hundred largest cities was 0.2 million in 1800 (Cohen, 2004: 27). It took a century and a half for the average size to reach the 2 million mark in 1950, while in another fifty years it increased by more than 3 times reaching a whopping 9.7 million by 2020 (Satterthwaite, 2020). In fact, the number of cities above any threshold has witnessed a dramatic increase after the onset of urban transition in the middle of the twentieth century. In 1800 there was just one city[1] in the world with a population of 1 million or more, and even after a century, i.e. by the turn of the twentieth century, the number of million plus cities could reach only 16. However, the number rose to 86 by 1950, 195 by 1975 and over 400 by the close of the last century (Cohen, 2004: 27). The latest available data indicate the presence of 548 million plus cities across the globe in 2018 (UN, 2019). It is estimated that by 2030, the world will add 158 new million plus cities in another twelve years (UN, 2019). Of these as many as 145 will be located in Asia, Africa and Latin America and the Caribbean while Europe and Northern America will add only 13 such cities. Rapid growth of cities is a recent phenomenon that has followed different trajectories in more developed and less developed parts of the world (Norton & Mercier, 2016: 388).

The distribution of the world's largest cities has undergone remarkable change with the fast emergence of a good number of the largest urban agglomerations in the less developed parts of the world. In 1950, there were only eight cities/urban agglomerations in the world having 5 million or more inhabitants of which five viz. New York, Tokyo, London, Paris and Rhein-Ruhr were located in high-income countries. The only urban agglomerations outside this region were Moscow, Shanghai and Buenos Aires. By the close of the century, these numbers increased to 32 and of them only 9 were located in the high-income countries. As per the latest available data there were 33 megacities—cities having 10 million inhabitants or more—in the world in 2018. Out of this, only five were located in the more developed regions of the world. London and Paris from Europe and New York and Los Angles from the United States that figured among the largest cities/agglomerations till the middle of the last century are now dwarfed, with respect to size of their populations, by emergence of huge cities in the less developed parts such Delhi, Shanghai, Sao Paulo, Mexican City, Cairo, Mumbai, etc. London and Paris even do not figure in the top 20 megacities of the world while New York, the largest urban agglomeration in 1950, and Los Angles occupy only 11th and 22nd positions in the list.

Another important contrast that exists between the current wave of urbanisation and the earlier experiences of more developed countries relates to the very nature of relationship between economic development and urbanisation. In the developed countries urbanisation was 'development induced' wherein 'pull factors' operating in urban areas played a key role. In other words, urbanisation was largely confined to countries or regions with high-income levels (Cohen, 2004: 34). As against this, the current wave of urbanisation is largely 'poverty induced'. The stagnant rural economy and resultant 'push factors' in low-income countries lead to migration to urban areas in search of better economic opportunities. Although migration is not necessarily the largest component[2] of urban growth, it does play a significant role in the overall process of urbanisation. In some parts of the world, for instance in Sub-Saharan Africa, towns and cities have grown without any improvement in economy and income levels.

## 1.4 Urbanisation in India's History

The Indian subcontinent shares a long history of urbanisation with Mesopotamia and the Nile Valley. Archaeological evidence indicates the presence of towns and cities in the Indus Valley region as early as the third millennium BC. However, as this was confined to the northern and north-western parts of the country, much of which lies now in Pakistan, the rest of the country remained untouched by urban life till the latter half of the first millennium BC. City life flourished in the Indus Valley region for over 600 years before coming to an end with the decline of the Harappan civilisation around 1700 BC. The decline of the Harrapan civilisation and the subsequent decay and final abandonment of the cities were followed by no urban development in the country for another thousand years or so. From 600 BC onwards, traces of towns and cities are again available from a more significant part of the country, both in the north and south. After that, the history of India's urbanisation is almost continuous, although the rise and decline of towns and cities also continued in the later phases.

The forthcoming paragraphs present a brief history of urbanisation, beginning with the Indus Valley civilisation through the ancient and medieval periods up to the end of British rule in 1947. The nature and process of urbanisation in the post-independence period are discussed elsewhere in the book.

**Pre-Historic Period:** Early historic period here refers to the period 2350–1800 BC. As noted already, the first wave of urbanisation on the Indian subcontinent dates back to the Harappan civilisation (sometimes called Indus Valley civilisation) four and a half millennia ago. Spread over a vast area stretching from Jammu in the north to South Gujarat in the south and from Baluchistan (a province in Pakistan) in the west to the western Uttar Pradesh in the east, Harappan civilisation witnessed flourishing urban life in several cities and towns of varying size and importance. Here in this region, we encounter a meticulous town planning system for the first time in human history. Cities were generally divided into two parts: a citadel, i.e. the higher ground, usually on the western side of the city with more eminent structures where the members of the ruling class lived, and the lower city where ordinary people lived. Houses were made of burnt bricks, and their arrangement over space followed a grid system. The roads and streets cut across one another almost at right angles, dividing the city into almost a series of rectangular or square blocks. For the discharge of wastewater, every city or town had an impressive drainage system.

Archaeological evidence indicates the presence of at least five major cities in the region. The two most important cities of the civilisation are Mohenjo-daro and Harappa, both now located in present-day Pakistan, on the banks of the rivers Indus and Ravi, respectively. Of the two, Mohenjo-daro is much older and spread over a larger area. The city is said to have been rebuilt several times without disturbing its original planned layout. On the other hand, Harappa was built as the second capital city at a later stage, perhaps with a purpose to accommodate the growing population. The city structure of both Mohenjo-daro and Harappa is identical. The other three major urban centres of the time were Dholavira, Ganweriwala and Rakhigarhi. Of them, two viz. Dholavira and Rakhigarhi are located in India (on the Rann of Kutch in Gujarat and in the Hisar district of Haryana, respectively), while Ganweriwala is located on the other side of India's boundary with Pakistan.

Besides, the ruins of several other lower-order cities or towns associated with the civilisation have been found within India's territory. The most important is Lothal on the Gulf of Cambay in present-day Gujarat. Lothal was a major port city then, providing trade links with the outside world. Apart from Lothal, other small port towns such as Suktagendor, Sotka-koh, Bhagatrav, Tuda and Somnath also facilitated trade at that time. Surkotada, Rangpur and Rojdi are other sites in Gujarat where ruins of urban centres of the time have been found. Archaeological remains of the Harappan city structure have also been found at Kalibangan in northern Rajasthan and Banwali in western Haryana. In addition, the ruins of over a hundred lesser sites associated with the civilisation have been discovered elsewhere within India's territory.

The population in these cities and towns was supported by well-developed agriculture in the hinterland. The major crops grown in the region were wheat, barley, peas, sesame, mustard, millet, rice, etc. In addition to a strong agricultural base, highly skilled craftsmen enabling mass production of crafts and a complex trade (including maritime trade with the outside world) were some of the factors underlying flourishing and sustained urbanisation in the region.

Opinions, however, differ on whether the emergence of the urban way of life in the region resulted from the diffusion of the Mesopotamian civilisation, possibly through migration, or was an indigenous development. Several arguments are put forward in support of the former hypothesis, which forms the basis of the proposition that there was an abrupt beginning of the urban way of life in the Indus Valley region. The emergence of an urban way of life in Mesopotamia precedes that in the Indus Valley region by at least a thousand years. Therefore, it is contended that the development may have come from Mesopotamia and the Nile Valley. Evidence of trade links between the two civilisations is also cited, supporting this idea. Sumerian coins have been found in Lothal, while Harappan coins have been found in Mesopotamia. Among others, similarities in the salient features of the two civilisations, for instance, the location of cities near river banks, wheel-turned pottery, the art of writing and the practice of agriculture, are also cited in support of this diffusion theory. Moreover, it is also sometimes claimed that archaeological evidence in the vertical diggings at the sites of Mohenjo-Daro and Harappa does not indicate any transition phase of evolution. However, a closer examination of the available evidence does not support the diffusion theory. Consider the following points:

1. Unlike Mesopotamia, where agriculture was based on a well-developed irrigation network, in the Indus Valley region, agriculture was purely a rain-fed system dependent on natural flooding of the plains. The Harappan people generally sowed the seeds in the flood plains after the floodwater had receded and reaped their harvest before the next flooding season.
2. The people in the region used harrows to till the land and were unaware of plough, which was extensively used in farming practices in Mesopotamia.
3. Unlike those in Mesopotamia, no defensive wall-like structures have been found around the cities in this region.

4. The design and artistic expressions on the pottery and other objects found in the ruins of the Indus Valley civilisation are undoubtedly of Indian origin. Its writing system is markedly different from those seen in Mesopotamia and elsewhere in West Asia.
5. The design of the coins and weights used in the Indus Valley civilisation is not similar to those of Mesopotamia.

Thus, it appears that the urban way of life was not transplanted in totality in the Indus Valley by migrants from Mesopotamia, although some ideas, for example, copper melting and the use of wheel-turned pottery, may have been borrowed from the latter (Ramachandran, 1989: 26). On the other hand, Sharma (2014) argues that Harappan urbanism must be seen as a result of bronze age culture with many similarities and differences outside Harappa.

The probable reasons for the decline of urban life in the Indus Valley may be purely man-induced or unleashed by some natural events or a combination of both. A gradual change in environmental conditions, such as a shift in climatic patterns and consequent agricultural disasters, perhaps resulting from excessive environmental stress caused by population growth and overexploitation of resources, invasions, brought about the end of cities.[3]

**Early Historical Period:** The first wave of urbanisation flourished in the north and north-western parts of the Indian subcontinent during 2350 BC to about 1750 BC. Once Harappan urbanism disappeared there was no urbanism in the country for more than a thousand years (Sharma, 2014: 11). Thus, no trace of urban development is found for a long period, at least up to the turn of the sixth century BC. Further, the urbanisation process was confined mainly to the Indus Valley region and did not penetrate beyond the upper Ganga plains. Down south, the entire peninsular uplands and the coastal plains also lacked urban culture. Evidence indicates the onset of the second wave of urbanisation on the Indian subcontinent only around the fifth to sixth century BC, in the Ganga plains in the north. It also marks the beginning of Northern Black Polished Ware, use of iron implements and the metal money. There was emergence of many urban centres in the Gangetic Valley during this time (Sharma, 2014: 11). This second wave of urbanisation was 'more durable and its effects are being felt to this day' (Erdosy, 1985: 82). It holds a particular significance in India's history as from this time onwards the process of urbanisation became a permanent feature of the cultural landscape of the country (Ramachandran, 1989: 34).

By the middle of the first millennium BCE, the Ganga plains became the home of several Aryan groups or tribes who fought incessantly among themselves and with the native people for control of resources. The clan identity thus gave way to territorial identity, leading to the emergence of a network of kingdoms. Sources indicate the presence of as many as eight *janapadas* or tribal territories in the Ganga plains. The territorial feuds among these kingdoms necessitated the development of capital cities and thus emerged some of the earliest cities of the country in this period. These cities included Hastinapur, Sravasti, Kapilavastu, Ujjain, Mahisamati, Champa, Rajgir, Ayodhya, Varanasi and Kausambi. The presence of silver and copper punch-marked coins and northern black polished ware indicate flourishing trade and commerce during that time. The river Ganga was the main trade route that connected the Ganga plains with the north-western parts of the Indian subcontinent. The incessant wars among these territorial units for supremacy resulted in the creation of new *janapadas*, some of which grew in size and prominence and were called *mahajanapadas*.

Buddhist writings and other sources indicate the presence of 16 *mahajanapadas*. One among them was Magadha. In the constant wars for supremacy and control of the Ganga Valley, the Magadha Empire finally emerged victorious around the middle of the sixth century BCE. Magadha, with its capital at Rajgrha (modern Rajgir), enjoyed an excellent geographical location. Pataligrama, which later came to be known as Pataliputra (modern Patna), has its origin during this period. The Magadha Empire had already extended its control over the eastern parts of the valley, providing them with rich revenue from the fertile lands and trade through rivers. In addition, access to coastal trade in the east also added to their strength. 'Neighbouring forests provided timber for building and elephants for the army. Above all, nearby rich iron ore deposits gave Magadha a lead in technology'.[4]

While the Gangetic Valley continued as the epicentre of urbanisation, with the ascendency of the Mauryan dynasty in 321 BC, the process further spread to the whole of North India by the third century B.C. and 'over Central India, Deccan and the Andhra region between the second century B.C. and third century A.D.' (Champakalakshmi, 1987: 68). While at its peak, the Mauryan dynasty ruled over almost the entire subcontinent (barring only areas lying south of the present-day Karnataka) and over much of present-day Afghanistan. The urban tradition reached its peak during this period. According to the Buddhist literature, apart from a hierarchy of administrative cities of which the capital city was the largest and most important of all, there were several centres of commerce at various levels, e.g. the *Putabhedana* (the largest of all specialising in wholesale trade), the *Nigama* (a market centre), the *Pattana* (a coastal trading town) and the *Dronamukha* (a town located at the mouth of a river that served as a port city). The cities of this period were rectangular or square in shape. The streets were of standard width of about eight metres and cut across each other almost at right angles, forming grids. The internal structure resembled the early Aryan villages and was marked with distinct levels of segregation based on occupation (Ramachandran, 1989: 40).

With the disintegration of the Mauryan dynasty towards the early second century BC and the rise of several small kingdoms in the north, some of the urban centres lost their prominence. However, the Ganga plains remained the epicentre of urban culture for roughly another five hundred years. With the ascendency of the Gupta dynasty in 320 AD, the political dominance of the Ganga plain in North India was once again restored. The Gupta dynasty ruled over the whole of North India, and its trade with the western and southeastern Asia continued unabated. Thus, city life flourished in the north and its adjoining regions in the Deccan, central India, and the Andhra regions. However, frequent invasions that began in the early fifth century AD from the northwest by the Hunas or Huns, a group from Central Asia, weakened the supremacy of the Gupta dynasty, and the trade links with the West were disrupted. Cities and, along with them, the urban way of life began to decline around this time, and by the middle of the sixth century AD, the Gupta dynasty came to an end, and northern India and part of central India were in the hands of the Huns. The rise of feudalism, a new socio-economic formation sometimes termed as 'land-grant economy' (Champakalakshmi, 1987: 68), further gave a severe blow to not only agriculture but also crafts and industries. With the decline of the empires and kingdoms and the rise of petty feudal chiefs, agriculture and industry became less productive due to which urban centres began to decline (Ramachandran, 1989: 43).

As noted already, overland trade links of the Ganga plains during the reign of the Mauryan dynasty played a significant role in the spread of urbanisation into the Deccan and Andhra region. In the extreme south, the region comprising modern-day Tamil Nadu and Kerala, urban way of life evolved independently with the emergence of Dravidian culture although evidence of trade links with the Mauryan Empire does exist. The literary evidence indicates the existence of cities in this part of the peninsula as far back as the fourth and third centuries BC. The major cities of that time were Madurai, Vanji, Musuri, Urayur, Puhar and Korkai. The Tamil region was divided into three major territorial units called *mandalams*, a term equivalent to *mahajanapadas* of the Ganga plains in the north. Each of them evolved in the fertile agricultural tracts of major rivers. The earliest towns or cities appeared in these fertile tracts of river basins or coastal plains.

Evidence indicates contact between the Tamil region and the north during Mauryan times. Trade with the Ganga plains in the north played a significant economic role. In addition, the region carried flourishing maritime trade links with western Asia, initially with southern Arabia and later with the Romans at the turn of the Christian era. With the decline in the Roman trade in the later phases, the region began maritime trade with the Southeast Asian regions. Unlike the northern plains, urban traditions in this part of the Indian peninsula exhibit a remarkable continuity since early historical times. One important reason for this continuity in urban tradition is the near absence of foreign invasions that characterise the northern plains (Ramachandran, 1989: 46).

**Medieval Period:** For the present purpose, medieval period begins with the end of the Gupta dynasty in the mid-sixth century AD and continues up to mid-eighteenth century AD when East India Company established its control. Spanning over a period of over one millennia medieval period, particularly its later half, is marked with significant revival of urbanisation in India. As already noted, urbanisation in the northern plains suffered a severe setback with the disintegration of the Gupta dynasty in the middle of the sixth century. Urban decay began in the fifth century when the Gupta dynasty was weakening. Several prominent cities of the earlier period were abandoned, although some could withstand the vagaries of history and maintain their glory. Several reasons are cited for the urban decay during the post-Gupta period. The first and foremost was political instability and frequent foreign invasions from the West. With the fall of kingdoms and the shifting of administrative seats, several prominent cities like Vaishali and Patliputra, to name a few, lost their glory. The invasion of Hunas, who entertained strong sentiment against Buddhism, led to the destruction of several cities associated with the Buddhists in both the north-western parts and the Ganga plains. In addition, frequent invasions and plunder by Mahmud of Ghazni in the early tenth century played havoc with the economic base and city life in the entire northern plains. Long-distance trade, the lifeline of urbanism, was disrupted, severely damaging the urban economy. There is also mention of natural calamities like famines, floods and earthquakes that adversely affected the region's agricultural base and city life. The introduction of the feudal system that destroyed the agricultural base, crafts and industries and the decline of Buddhism in the post-Mauryan period are also important reasons for urban decay in the post-Gupta period.

A distinctly slow pace of urbanisation, thus, characterises the early medieval period due to political instability and other associated reasons. However, the establishment of Muslim rule, i.e. the Delhi Sultanate, led to a significant revival of urbanisation. The Delhi Sultanate—beginning with the Slave dynasty in the year 1206 AD and followed by the Khilji dynasty, the Tughlaq dynasty, the Sayyid dynasty and the Lodi dynasty—ruled over a large part of the Indian subcontinent for more than three hundred years. During the rule of these dynasties, urbanisation spread to almost the whole of northern India and its adjoining regions. Delhi and Agra, which served as capital cities at different points under the Delhi Sultanate, were the two prominent cities of northern India. The other major cities were Mathura, Thaneshwar, Allahabad (present-day Prayagraj), Varanasi, Pataliputra, Gwalior, Ujjain, Dhar, Somnath, Meerut, Panipat, Bharuch, Baroda (present-day Vadodra) and Srinagar, among others (Ramachandran, 1989: 51).

Contrary to this, the experiences of the extreme southern parts of the Indian peninsula were relatively different. Urbanisation and city life flourished in the south with the emergence of powerful Hindu kingdoms. During the early medieval period cities and towns grew in number many of which are still well known today. Several temple towns came into existence under the patronage of Hindu kings with the spread of Puranic religion and rise of the Pallavas and the Pandyas. Srirangam, an island city by the Kaveri River, and Thanjavur, one of the largest cities of present-day Tamil Nadu, are among these temple towns. The expansion of agriculture that took place during the seventh and eighth centuries provided an impetus to commercial activities. The merchant organisations or guilds also played a key role in promoting trade and commerce. With the revival of Asian trade, several new port towns, such as Nagappattinam (at the mouth of the Kaveri), Kovalam (on the Kerala coast to the south of Thiruvananthapuram), Cuddalore (in Tamil Nadu), etc., are some of these port cities or towns that came into existence during the time.

The Mughal period stands out as the second high watermark of urbanisation in India's history, the first occurring during the Maryan period. The Mughals established their rule in 1526 AD and had an effective control over much of the Indian subcontinent up to the middle of the seventeenth century AD. At their peak, they ruled over a vast area stretching from Afghanistan and Pakistan in the northwest and Gujarat in the west to Assam and Bangladesh in the east, and from Kashmir in the north to uplands of the Deccan plateaus in the peninsula. Although the Mughal dynasty continued to exist after the middle of the seventeenth century AD also, their influence and power had considerably reduced, and by the mid-nineteenth century, the Mughal dynasty came to an end. Political stability that lasted for over two centuries provided suitable conditions for the growth of towns and cities. With the administrative reforms introduced by Akbar, the empire was divided into 15 *subahs* or provinces which were further divided into more than hundred districts known as *sarkars* further divided into *mahals* or *parganas*. Each of the provinces housed several cities and towns. The absolute size of urban population in these cities and towns was larger than the urban population in Europe, and anywhere between 10 and 15 per cent of the total population in the Mughal empire resided in urban areas. Abul Fazl, a courtier of Akbar, mentions the presence of as many as 2, 837 towns in the Mughal empire towards the close of the sixteenth century AD.

Spread of urbanism during the Mughal period owes more to the revival of older established cities than development of new ones. There were indeed very few new cities, e.g. Moradabad and Fatehpur Sikri that were developed during the period. While the former still exists today, the latter is in ruins now. In the year 1571 AD, Emperor Akbar built Fatehpur Sikri as the capital city but it served as capital city for barely 15 years and the capital was shifted back to Agra. The city was completely abandoned in the early seventeenth century. The urban landscape during the Mughal period was dominated by some 15 to 20 large cities which served as administrative seats as well as centres of industrial and commercial activities. Many of them not only survive to this day but occupy very high positions in urban hierarchy at the present time. Mention may be made of Delhi, Ahmadabad, Jaipur, Lucknow, Patna, Srinagar, Varanasi, Allahabad and Agra, all million plus cities/agglomerations at the time of last census that was held in 2011. Among other prominent cities of that time are Ajmer, Ujjain, Awadh, Jaunpur, Burhanpur, etc.

Urbanisation during the Mughal period was marked with a hierarchy of urban centres of which the capital cities being the most important ones occupied the top position. The Mughals had multiple capitals and the administrative seat shifted back and forth between these centres. Delhi served as the initial capital of the Mughals. Later the capital was shifted to Agra which was already an important city before the ascendency of the Mughals. Well connected with Kandahar and Kabul via Lahore, Agra soon became an important hub of industries and commerce. During the reign of Akbar, a new city known as Fatehpur Sikri was developed and the capital was shifted from Agra to Fatehpur Sikri in 1571 AD. However, the geographical location of the city was not suitable and hence it served as the seat of the Mughal Empire for hardly a decade and a half. Later in 1585, the capital was shifted to Lahore by Akbar perhaps due to some political and military concerns. In 1598, the capital was shifted back to Agra which remained the seat of power for another 50 years. Subsequently, the emperor Shahjahan built a new planned city called Shahjahanabad, a walled city, in the vicinity of present-day Delhi and the capital was once again shifted from Agra to Delhi. Since then Delhi has occupied a very prominent place in the urban landscape of the country. Emperor Shahjahan also built an ivory-white marble mausoleum the *Taj Mahal* that houses the tomb of his wife, near Agra on the right bank of the river Yamuna outside the city proper. One of the most beautiful monuments of the world, the Taj Mahal continues to attract tourists from around the world providing an important source of income to the government. In addition, the city continues as an important centre of industrial and commercial activities in north India to this day. Other major cities included centres of administration at various levels as well as commerce and industries. On the other extreme, there was large number of small towns scattered in different parts of the territory under the Mughals. These towns flourished under the support of officials and feudal chiefs who received land grants from the Mughal emperors. The petty feudal lords helped in the process of land resettlement and the building of small towns (Ramachandran, 1989: 53). The land-grant system which was in practice from the time of the Delhi Sultanate, encouraged labour classes and artisans to settle in these towns and carry their professions. Thus, the system not only promoted urbanisation but also helped the rulers in connecting with rural masses through these intermediate functionaries.

The expansion of traditional industries, including textiles, metal works, arts and crafts and flourishing trade (both external and inland), played a crucial role in urbanisation during the Mughal period. The Mughals, particularly during the time of Akbar, took specific measures to promote indigenous or traditional industries. According to Abul Fazl, Akbar maintained hundreds of imperial factories, each resembling a large urban centre and producing goods for royal consumption (Iftikhar, 2018: 70). These centres provided employment to a large number of workers. Indian textiles, including cotton, silk and woollen, were famous outside the Indian subcontinent. Cities across the plains, particularly in the Bengal plains, developed into specialised fine cotton textile production centres. Dacca, present-day Dhaka, the then capital of Bengal province under the Mughals, was famous for its delicate cotton fabrics known as *muslin*. In addition, carpet making, pottery, leather works and wooden products also flourished in the cities across the plains. Chittagong, now a large port city in Bangladesh located on its southeastern coast, specialised in shipbuilding. The ships built at Chittagong were of high standard and were supplied to even distant places like Istanbul in West Asia. Indian products were in high demand in the markets of West and Southeast Asia as well as Europe, which promoted trade and commerce. Many cities on the fringe of the Mughal empire, including coastal areas, emerged as important centres of commerce and trade. Cambay, Surat, Burhanpur, Satgaon, Chittagong and Hoogly are some of these important centres. Burhanpur, situated on the north bank of the river Tapti in present-day Madhya Pradesh, was a significant centre of trade and commerce between the Mughal empire and the kingdoms of the Deccan. All the major urban centres of that time, some of which now lie outside India's territorial jurisdiction in Pakistan and Bangladesh (e.g. Lahore, Multan, Sialkot, Dacca and Chittagong, to name a few), were invariably the important centres of trade and commerce. These urban centres were densely populated with large market to cater to the needs of the local population. They also served as collection centres of goods for both inland and external trade. Smaller towns, with their markets, served the daily necessities of local inhabitants and people from the surrounding villages.

In the late seventeenth century the Mughals had extended their control over much of the Indian peninsula in the south, barring only the extreme south. Much of this expansion was driven by India's growing commercial and cultural contact with the outside world. Growth and expansion of urbanism that had occurred earlier during the regimes of the Bahmani Kingdoms and Vijaynagar Empire (and also the Marathas) received further momentum after the region fell under the control of the Mughals. Cities like Pune, Bijapur, Ahmadnagar, Gulaberga, Kolhapur, Hyderabad, Badami, Kolhapur, Golkunda and Hampi that flourished during the earlier regimes continued to grow. While some of them, like Golkunda and Hampi, are in ruins today, many continue to dominate the urban landscape of the Indian subcontinent to this day. Hyderabad and Pune agglomerations are now among the megacities of India, with more than 5 million inhabitants each. As already mentioned earlier, in the extreme southern parts of the Indian peninsula, the growth of urban centres had flourished in the form of the 'temple towns' under the patronage of various Hindu kingdoms (i.e. the Cholas, the Pandyas, the Western Gangas) during the early medieval period and beyond (Champakalakshmi, 2014). By the fourteenth century AD, these kingdoms ended one after another and, thus, followed a period of stagnant urban growth.

The Mughal empire reached its height during Aurangzeb's reign (1658–1707 AD). However, his political and religious intolerance and discriminatory policies resulted in rebellions in different parts of the empire. The empire began disintegrating towards the middle of the eighteenth century, and its rule was reduced to a small area around Delhi, which finally ended in 1857.

**Colonial Period:** The arrival of Europeans in India through the sea route via the Cape of Good Hope for trade, beginning in the early years of the sixteenth century, was a turning point in India's history. The first to arrive were the Portuguese. Vasco da Gama, a Portuguese navigator, discovered this sea route to India in 1498. The Portuguese held a monopoly over trade through the Indian Ocean for almost a century before the Dutch, the French, and the British arrived in the seventeenth century. They developed trade posts or port towns on the eastern and western coasts. Indian goods were much sought after in Europe, so trade with the Indian subcontinent meant colossal wealth for these countries. Rivalry among these European colonial powers was inevitable. It led to constant conflict and wars, in which the British East India Company finally emerged victorious. They had established their base in Madras (now Chennai) in 1639 and Calcutta (now Kolkata) in 1690. The British soon started interfering in local politics, and their role in, and outcome of, the battle of Plassey in 1757 marked the beginning of British imperialism in the Indian subcontinent. Soon, the company realised they could expand and consolidate their power by pitching one small kingdom against the other. By the middle of the nineteenth century, they had established their control over not only the Indian subcontinent but also parts of Southeast Asia. After the revolt of 1857, the East India Company was abolished, and India was subject to the direct rule of the British Crown. From then onwards, till 1947, when India gained independence, Britain exercised complete control over the entire subcontinent, bringing the entire country under one umbrella, an unprecedented development. The British colonial economic policies, thus, governed the course of urbanisation in the country for almost two hundred years.

The immediate effect of the British ascendency in India was disastrous to India's economy and urbanisation. The latter half of the eighteenth century saw a series of revolutionary changes in the production technology in Britain. The economy of Britain, hitherto based on agriculture and handicrafts, was replaced by factory-based production on a large scale. Britain almost monopolised world trade in the years that immediately followed the Industrial Revolution. India soon became the market for finished goods from Britain and the source of raw materials, leading to the systematic destruction of India's traditional industries and crafts, which were the lifeline of urban centres during the pre-British period. The traditional pattern of trade and commerce was disrupted. A new stage of industrial capitalist exploitation thus set in with the transformation of India's economy to a colonial market. The worst hit was India's textile industry, with the emergence of Manchester as a world centre of large-scale cheap cotton textile production. Back in India, the livelihood of millions of artisans and craftsmen was destroyed, and cities and towns, which earlier enjoyed eminent positions in producing goods and trade within and outside the subcontinent for ages, soon began declining. Cities like Dacca, Murshidabad, Patna, Cambay, Vijaynagar, Lucknow, Ahmedabad and Surat, among many others, soon lost their glory. According to an estimate, in the year 1800 AD, approximately 11 per cent of India's population lived in urban areas, which came down to 8.7 per cent at the time of the first census in 1872 (Ramachandran, 1989: 61). 'Soon, the regional structure of the economy began changing, and there emerged a new urban landscape dominated by port cities that facilitated exploitative trade and commerce of the British' (Hassan, 2020: 56).

The colonial rule radically disrupted the urban processes in the country and insulated it from the pre-colonial urban tradition. The major cities of nineteenth- and twentieth-century India were the creations of colonialism not integrated with the existing urban hierarchy of the medieval urban system (Bonner, 2004: 2016). Kolkata (erstwhile Calcutta), established by the East India Company in 1690 as a trading post, had already emerged as the country's largest city[5] by 1872 and one of the largest cities of the entire British Empire. The other two port cities of Bombay and Madras established during colonial rule (now Mumbai and Chennai, respectively) also witnessed phenomenal growth in their population. These port cities were the leading administrative, industrial and commercial cities of British India. The decline in indigenous industries and impoverishment of the rural economy in the wake of faulty agricultural policies resulted in stark regional inequality. Thus, a large-scale population migration from the hinterland, both countryside and small towns, to the port cities in search of employment began, leading to unprecedented growth in the population of the port cities.

In contrast, most of the prominent cities of the pre-British period were either stagnant or declining. The decline of pre-British cities and towns was also facilitated by the introduction of the railway network, which began in 1853. Within a decade, all three port cities were already connected by railheads with strategic places in their hinterlands to facilitate the export of raw materials. By the end of the nineteenth century, the rail network had covered almost the whole country (total length of 40,000 kms), and every railway station became a centre of export of raw materials. Although the introduction of railways later saw the emergence of several new industrial cities like Jamshedpur, Asansol and Dhanbad, among many others, it replaced the traditional trade channels, thereby adding to the woes of the earlier centres of trade and exchange. The interactive system that had previously evolved through ages between centres of handicraft or service or trade and commerce, on the one hand, and their hinterland of primary production, on the other, as well as the one between large cities and smaller towns over space, was the main casualty (Kundu, 2011: 3). The emerging urban landscape was characterised by a hierarchy of newly established urban centres that served as satellites of port cities. 'The medieval urban system of India, which comprised several sub-systems that had evolved around regional cores, was replaced during the colonial period by a pattern characterized by the emergence of a few dominant cores and the decline and decay of provincial cities' (Habeeb, 1987: 11). A high degree of primacy thus became an integral feature of India's urbanisation in the years to come.

Notwithstanding these initial negative impacts, British economic policies and governance promoted urbanisation in several ways. New cities and towns came up all over the subcontinent due to the British's deliberate efforts or the indirect effects of building infrastructure, including the railway network. The capital city of the British Empire was shifted from Kolkata to New Delhi in 1911. New Delhi was a completely new planned city with magnificent buildings housing administrative and commercial complexes, wide roads, modern amenities and greeneries near Shahjahanabad, the capital city of the Mughals. As part of present-day Delhi agglomeration, the city witnessed tremendous demographic expansion since its inception and is currently one of the world's largest agglomerations. Although much of the growth in the population of the city occurred during the post-independence period, Delhi, along with the three port cities of Kolkata, Mumbai and Chennai, dominated the urban landscape of colonial India. Guided by their trading interests, the British developed several other port cities along the western and eastern coasts. The spread of rail networks led to the development of industrial towns including Jamshedpur known for its iron and steel industry under Tata group. Tata steel plant later grew into the largest single steel complex in the British Commonwealth by World War II. Several other new towns that primarily served as railway stations across the length and breadth of the country came into existence. Initially, these railway stations were mainly the collection points of goods for export. However, in due course of time, with the development of railway colonies housing the rail staffers and the location of commercial establishments accompanied by the growth of other basic amenities, these places acquired important positions in the urban hierarchy, providing essential linkages with the surrounding small towns and the countryside. Several large villages near these railheads gradually developed into small towns. The development of the rail network not only facilitated trade activities but also led to the emergence of several industrial towns, thereby providing a new boost to the urban processes.

The urban structure of many pre-British cities and towns was subject to modification during the period. The modification process was more noticeable in places that served as centres of administration at various levels, e.g. the imperial capital, the provincial capitals and the district and tehsil headquarters. In the centres of higher-level administration, a portion of the city was developed as the *Civil Lines*, which accommodated administrative offices, courts and residents of officials. Based on the plan, the civil lines were characterised by low population density and horizontal development, sprawling official bungalows, wide roads and streets, open spaces and sufficient greenery. In contrast, the central city represented crowded agglomerations of houses with extremely high population density and poor quality of life. Likewise, *cantonments* were developed in strategic locations near some large cities to accommodate military personnel. However, separate areas within the cantonments were devoted to military personnel of British origin and Indian soldiers. More than a hundred such cantonments were developed during the British Raj, and they were located in the north and north-western parts of the subcontinent. The quality of life regarding basic facilities was invariably better than that of the city proper.

Another significant contribution of the British was the development of a chain of *hill stations* in different parts of the country. Primarily built to provide a respite to the British officials of the main administrative centres from the scorchy heat of the summer season, these places soon became eminent urban centres. Although significant changes have taken place in their landscape after independence, these hill stations continue to play an important role in the economy of the surrounding areas even today. Shimla, Mussoorie and Nainital in the north near Delhi, Darjeeling and Shillong in the east near Kolkata, Mahabaleshwar in the Western Ghats near Bombay, and Nilgiri and Kodaikanal area near Chennai are some of the glaring examples of such hill stations. Along with the creation of hill stations, the introduction of tea and coffee *plantations* also gave rise to the emergence of settlements with distinct urban characteristics and contributed to urbanisation at that time. British tea plantation started first in Assam hills in the 1850s in the north-eastern region, and soon spread to the Darjeeling hills of north Bengal. Tea plantation started in the Nilgiri Hills of Tamil Nadu some 15 to 20 years later. Much of the tea production was exported to Britain. Coffee plantation was also started in the southern states of Tamil Nadu and Kerala towards the middle of the nineteenth century. Although these plantation settlements were generally small in size, they contained residential houses of the workers, a market place for daily needs, schools, and health-care centres supported by other basic amenities.

Despite the unquestioned control of the British Raj for almost a century, a substantial part of the Indian subcontinent remained outside the direct control of the British. They represented the princely states scattered in different parts of the subcontinent, the number of which ran into hundreds,[6] some of which were as large as the British provinces. At the time of partition in 1947, these princely states accounted for as much as four-tenths of the geographic area of pre-independent India and one-fourth of its population. These princely states had their own governance, and hence the growth and development of urban centres in these princely states were largely independent of the British economic policies. The only direct British impact on urban processes was seen in the institution of Residency and cantonments in the capital cities of some of these princely states (Ramachandran, 1989: 68). The office of the Residency was a political instrument that dealt with the relations between the Government of India and the princely states (Copland, 2004: 807). Thus, the Residency and the adjoining areas of these princely states experienced rapid growth in contrast to other small towns that remained almost stagnant. The major cities of these princely states, mainly the state capitals, which received special attention from the ruling class, flourished during the period.

During the British rule, enumeration of population and mapping of the territory were important administrative activities undertaken by Census of India and Survey of India, respectively. The rural and urban classification was attempted since the first synchronous census of 1881, a definition of urban evolved and was debated leading to the measurement and analysis of the degree of urbanisation and its relationship with economic and social processes (Bose, 1964). It was the beginning of statistical urbanisation which has changed the nature and practice of urban studies in India.

## Notes

1. While quoting Chandler (1987), Cohen writes that erstwhile Peking was the only million plus city in the world in 1800 (Cohen 2004: 27), Satterthwaite mentions that London topped the largest 100 list with 1.1 million inhabitants.
2. Natural increase, i.e. the gap between births and deaths in urban areas, net migration and reclassification that includes change in the civic status of settlements as well as expansion in municipal limits are the components of urban growth.
3. Britannica, *The end of the Indus Civilisation*, https://www.britannica.com/place/India/The-end-of-the-Indus-civilization.
4. Britannica, The beginning of the historical period, c. 500–150 BCE, https://www.britannica.com/place/India/The-beginning-of-the-historical-period-c-500-150-bce.
5. Kolkata had grown into an agglomeration of more than 1 million inhabitants as early as the time of the first census enumeration in 1872 (Venkatarangan, 1989: 315).
6. At the time of independence there were about 600 such princely states in different parts of the country, https://oxfordre.com/asianhistory/view/10.1093/acrefore/9780190277727.001.0001/acrefore-9780190277727-e-355.

## References

Bonner, P. L. (2004). Migration, Urbanization and Urban Social Movements in Twentieth Century in India and South Africa. *Studies in History, 20*(2), 215–236.

Bose, A. (1964). A Note on the Definition of 'Town' in the Indian Census: 1901–1961. *The Indian Economic and Social History Review, 1*(3), 84–94.

Champakalakshmi, R. (1987). Urbanisation in South India: The Role of Ideology and Polity. *Social Scientist, 15*(8/9), 67–117. https://www.jstor.org/stable/3520287. Accessed on 25 February 2024.

Champakalakshmi, R. (2014). Urban Processes in Early Medieval Tamil Nadu. In I. Banga (Ed.), *The City in Indian History: Urban Demography, Society and Politics* (pp. 47–68). Manohar.

Chandler, T. (1987). *Four Thousand Years of Urban Growth: A Historical Census*. St. David's University Press.

Cohen, B. (2004). Urban Growth in Developing Countries: A Review of Current Trends and a Caution Regarding Existing Forecasts. *World Development, 32*(1), 23–51.

Copland, I. (2004). Princely States and the Raj. *Economic and Political Weekly, 39*(8), 807–809.

Davis, K. (1955). The Origin and Growth of Urbanization in the World. *American Journal of Sociology, 60*(5), 429–437. https://www.jstor.org/stable/2772530

Erdosy, G. (1985). The Origin of Cities in the Ganges Valley. *Journal of the Economic and Social History of the Orient, 28*(1), 81–109. https://www.jstor.org/stable/3631864. Accessed 16 February 2024.

Habeeb, A. (1987). Problems of Urbanization in India Prior to 1947. In S. M. Alam & F. Alikhan (Eds.), *Perspectives on Urbanization and Migration: India and USSR*. Allied Publishers Private Limited.

Hassan, M. I. (2020). *Population Geography: A Systematic Exposition*. Routledge.

Iftikhar, R. (2018). Urban Formation and Culture Transformation in Mughal India. *South Asian Studies, A Research Journal of South Asian Studies, 33*(1), 67–83.

Kundu, A. (2011). *Trends and Processes of Urbanisation in India., Urbanisation and Emerging Population Issues-6*. Human Settlement Group, IIED, Population and Development Branch, UNFPA.

Lowry, I. S. (1990). World Urbanization in Perspective. *Population and Development Review, 16*, 148–176.

Murphey, J. C. (1970). Literacy. In *Encyclopedia Americana* (vol. 14, pp. 775–777). Americana Corporation.

Norton, W., & Mercier, M. (2016). *Human Geography* (9th ed.). Oxford University Press.

Ramachandran, R. (1989). *Urbanization and Urban Systems in India*. Oxford University Press.

Satterthwaite, D. (2020). The World's 100 Largest Cities from 1800 to 2020, and Beyond. International Institute for Environment and Development, Blog. https://www.iied.org/worlds-100-largest-cities-1800-2020-beyond. Accessed on 15 February 2024.

Sharma, R. S. (2005). *India's Ancient Past*. Oxford University Press.

Sharma, R. S. (2014). Urbanism in Early Historic India. In I. Banga (Ed.), *The City in Indian History: Urban Demography, Society and Politics* (pp. 9–18). Manohar.

Tellier, L.-N. (2019). *World Urban History: An Economic and Geographical Perspective* (2nd ed.). Springer Nature Switzerland.

UN. (2019). *World Urbanization Prospects, The 2018 Revisions*, Department of Economic and Social Affairs, United Nations, New York.

Venkatarangan, L.B. (1989). Population Trends of First Four Megacities in India. In S.N. Singh, M.K. Premi, P.S. Bhatia & Ashish Bose (Eds.) *Population Transition in India* (pp. 309–318). B.R. Publishing Corporation.

# Chapter 2
# On Defining Urban and Rural–Urban Classification

## 2.1 Introduction

The nature of urbanisation is premised on how we define urban. Ontologically, just as the concepts of modernity or development, urban is also a debatable concept that entails diversity in terms of criteria defining urban. Nevertheless, most of the countries have evolved their definition of urban, and developed a framework for rural–urban classification for collection and compilation of data. It is, however, worth mentioning that in almost all the countries, including India, urban is specifically defined, and the rest is treated as rural. This has been challenged by many researchers who instead recommend a 'rural–urban continuum approach' (see, for instance, McGee, 1991; Hugo et al., 2003). However, as a matter of convenience, most of the nations continue with the dichotomous approach in the categorisation of settlements into rural and urban keeping the methodology simple and comparable that enables analysis based on administrative and municipal boundaries.

The classification of settlements into 'rural' and 'urban' is relatively a development of the recent past. The onset of Industrial Revolution in Europe in the mid-eighteenth century and the resultant growth in infrastructure and services in certain segments of space provided a distinct character to a settlement or a group of settlements that later began to be recognised as towns or cities. The concentration of population, predominance of non-agricultural activities and the provision of social amenities emerged as distinguishing features of cities and towns following the industrialisation of agrarian economies (Bhagat, 2005).

United Nations (2019a) defines urbanisation as a 'complex socio-economic process that transforms the built environment, converting formerly rural into urban settlements, while also shifting the spatial distribution of a population from rural to urban areas. It includes changes in dominant occupations, lifestyle, culture and behaviour, and thus alters the demographic and social structure of both urban and rural areas' (United Nations, 2019a: 10).



R. B. Bhagat and M. I. Hassan, *Urbanisation and Urban Policies in India*, Exploring Urban Change in South Asia, https://doi.org/10.1007/978-981-96-3181-0_2

The nature of governance in urban areas soon differed from that in rural areas. With the increasing complexity of urban functions like sanitation, water supply, waste collection and disposal, control of infectious diseases, education, roads and street lights, parks and recreational spaces, markets of various goods, etc., town planning was part of the agenda and activities of urban governance. Thus, urban and rural local bodies came into existence in different parts of the world. In India, the rural local bodies are known as *Gram Panchayat* (i.e. village council), and urban local bodies, are known by different names such as Municipal Corporation, Municipality, Municipal Council, *Nagar Panchayat* (i.e. Town Council), Notified Area Committee, etc., depending upon the size of urban population.

Rural–urban classification provides a significant basis in the formulation of development planning and serves as required categories for local governance. While it is expected that rural and urban categorisation should precede the accorded status of local governance and administration, it is also found that the administrative and political reorganisation of settlements define their rural and urban status. As rural and urban areas differ markedly from each other in terms of their social, demographic and economic characteristics, this necessitates different forms of governance and planning processes for the two. Further, in principles, rural–urban classification may facilitate decentralised governance, and promote self-reliance, competitiveness and efficiency. It may also provide a perspective of development from below which is local in fulfilling the aspirations of the people.

Thus, rural–urban classification has emerged as an instrument of political empowerment through decentralisation closely associated with the level of development of the settlements, governance and allocation of funds (Reza, 2002). As such, there has been an increasing concern regarding the improvement in the parameters used in identification and delimitation of urban areas among the researchers as well as national agencies engaged in rural–urban classification, defining urbanisation and demarcating cities and towns (Champion & Hugo, 2003; Halfacree, 1996; Pandey, 2003; United Nations, 2019a; Wirth, 1938).

A scrutiny of the rural–urban classification in different parts of the world reveals expanding industrial activities as the main force behind the process of urbanisation. As against this the predominance of agricultural activities characterises the rural areas. However, the initial 'agriculture-industry' dichotomy gave way to 'agriculture-non-agriculture' dichotomy in later times. Along with economic criteria, rural–urban classification is also influenced by administrative, political, demographic and infrastructural criteria. In short, rural–urban classification is about classification of settlements; it is about where people live and work. In other words, it is about people, space and place relationships, and their dynamics shaped by economic, demographic and political changes.

## 2.2 Rural–Urban Classification

Rural and urban classification in different countries is attempted using a set of quantitative indicators. The size and density of population, proportion workforce outside agriculture, administrative status of the place, etc.,[1] in varying degrees, are the most commonly used indicators in different countries to define urban. The United Nations Population Division updates the criteria of rural–urban classification as adopted by different countries and publishes the data on the trend and prospects of world urbanisation biennially. For example, data compiled for 233 countries or areas by the United Nations Population Division in its 2018 round of revision shows that 121 countries adopted administrative criteria to distinguish between urban and rural areas. Among these, 59 countries used administrative designations as the sole criterion. The criterion of population size or population density was used in 108 countries, and 37 countries used demographic characteristics as the sole criterion. It is very interesting to observe that the lower limit above which a settlement is considered to be urban varied considerably, ranging between 200 (Iceland, Greenland, Sweden and Denmark) and 50,000 inhabitants (Japan). Economic characteristics were part of the criteria used to identify urban areas in 38 countries or areas. However, economic characteristics were not used as the sole criterion in any of the countries. Certain characteristics, such as the existence of paved streets, water supply systems, sewerage systems, or electric lighting, were part of the definition of urban in 69 countries, but only in eight cases were such criteria used alone. There were 12 countries or areas that did not use any definition or used an unclear definition of what constitutes urban areas and in another 12 cases the entire population of a country or area was considered to be urban (United Nations, 2019a).

Since the 1990s, in India, local bodies were rediscovered emphasising a philosophy of new governance fundamentally different from the role of a conventional government. It included not only the organisation of and relationships between political and administrative institutions but also the relationships among government, private institutions and civil society. It represents the relationship between the government and the governed encompassing issues of accountability and empowerment, particularly of those marginalised (IIPO, 2001; UN-Habitat, 2002). Keeping in view the new philosophy of governance, the Indian Parliament amended the constitution known as the 73rd and 74th Amendments in the early 1990s. These amendments granted constitutional status to the rural and urban local bodies in the federal structure of the country, and provided for their autonomy in preparation and execution of development plans and formulation of measures for generating their own resources besides grants from the state governments. The constitutional provisions and legislative framework initiated over the decades have given local bodies a legitimate role in the federal structure of the country. Thus, it is important to understand the criteria of defining urban in India and other countries of the world. This chapter demonstrates that rural and urban classification is highly significant for local governance and development planning in the Indian context.

As will be seen later in the chapter, urban centres in India are classified into two types—statutory towns and census towns. The statutory towns are served by Municipal Corporation, Municipality, Notified Area Committee, Town/*Nagar Panchayats*, etc., depending upon the size of the population. The respective state governments grant the municipal status to the urban centres. As against this, the census towns which are identified by the Census of India—an organ of the central government, are mostly governed by *Gram Panchayat* (Village Council). The dynamics boundaries between rural and urban areas are not co-terminus with municipal and non-municipal boundaries, and, therefore, what is urban may not be necessarily municipal. The dissuasion in the forthcoming section revolves around the convergence and divergence between the processes of urbanisation and municipalisation.

As noted already, while it is the Census of India which defines and classifies settlements into rural and urban areas, the state governments grant municipal status to urban centres within their respective jurisdictions. In the light of this, a critical examination of the criteria adopted by the census organisation for rural–urban classification and its congruence with the dynamics of state-accorded municipal/non-municipal status becomes essential. The present chapter addresses this issue also in the context of its underlying implications for municipal governance and urban development in the country.

In South Asia, while in Nepal any place with a population above 9000 is treated as urban, in other neighbouring countries like Bangladesh, Sri Lanka and Pakistan only administrative criteria are taken into account in rural–urban classification. Thus, all settlements having Municipal Corporation, Municipality, Town Committee, Urban Councils, etc., are treated as towns/cities (United Nations, 2019a).

As against this, in China, rural–urban classification is to a large extent political. Urban areas are defined as constituting town and its countryside together based on the extension of town government seats. People engaged in non-agricultural pursuits in the areas of town government seats enjoy certain benefits that are not available to others. Thus, rural–urban classification in China is characterised by differential privilege to urban dwellers (Zhu, 2001). The administrative designation forms a very important criterion in the official definition of urban population and urban settlements in China. There are three administrative types of cities in China: *first*, provincial-level municipalities (Shanghai, Beijing, Tianjin, Chongqing); *Second,* prefecture-level cities (PLCs); and *third* county-level cities (CLCs). Both PLC and CLC designations are based on population, economic, public finance and infrastructure criteria. In addition, a *fourth* category of administrative towns with a population of less than 100 thousand is also considered as urban settlements. One of the main drawbacks of the administrative criteria is that the population base used in calculations excludes migrants and those lacking urban *hukou*[2] (a household registration system based on residency status) while many other countries do not differentiate between migrants and non-migrants in defining urban populations. However, since the 2000 census enumerated populations where they actually resided at the time of the census, if they had maintained residence for longer than six months. Apart from administrative status, China also adopted a statistical definition based on some important spatial characteristics namely urban density (at least 1500 persons per square kilometre) and contiguity of urban constructions such as having contiguous built-up area, being the location of the local government, or being a street (*jiedao*) or having a resident committee. In addition, the residents living in villages or towns in outer urban and suburban areas that are directly connected to municipal infrastructure and that receive public services from urban municipalities are also defined as urban population (OECD, 2010; United Nations, 2019a). Thus, the definition of 'urban' has been widened recently in China and it classifies a vast number of places under rural administration as urban and counts the residents as urban population what is described in urban literature as in situ urbanisation, i.e. place-based urbanisation without induced by rural to urban migration leading to the transformation of rural areas to include urban features, infrastructure and services. In both China and India this significant form of urbanisation poses a serious challenge in rural–urban classification (Bhagat, 2016; Zhu, 2017).

In the United States well above four-fifths of the population resides in urban areas (UN, 2019b). The rural and urban categories continue to hold their significance with respect to allocation of funds for social sectors such as education and health services (Chen et al., 2019; Congress of the US, 1989; Huang, 1995). More importantly the definition of urban has also been changing over the censuses. According to the 2020 census 'an urban area will comprise a densely developed core of census blocks that meet minimum housing unit density requirements, along with adjacent territory containing non-residential urban land uses as well as other lower density territory included to link outlying densely settled territory with the densely settled core. To qualify as an urban area, the territory identified according to the criteria must encompass at least 2000 housing units or at least 5000 persons. The term "rural" encompasses all population, housing, and territory not included within an urban area' (Census Bureau, 2022). Within urban, it identifies an urban nucleus with a housing density of 1275 housing units per square mile, initial core of housing density of 475 housing units per square mile and 200 for non-continuous territory to expand the urban block agglomerations into less dense, but structurally connected portions of urban areas (Census Bureau, 2022). It is important to note that rural–urban classification in the United States is attempted *post facto,* i.e. after the results of census counts are available whereas in India rural–urban frame is prepared prior to the actual census based on information from the preceding census (Bhagat, 2005; US Bureau of Census, 2001).

In Europe, for instance, countries like the United Kingdom and Sweden define urban area on the basis of built-up area. Thus, in the United Kingdom, a continuously built-up area with at least 10,000 persons is defined as urban (United Nations, 2019a). The largest urban areas are defined as agglomeration divided into various urban sub-divisions. Similarly, in Sweden, a continuously built-up area with at least 200 inhabitants is defined as urban provided houses are located not more than 200 m apart from each other. This definition has not changed until recently (UN, 2001; United Nations, 2019a). In Brazil too, urban areas are 'legally defined on the basis of buildings, streets, and intense human occupation; this includes areas that have been affected by transformations resulting from urban development and those reserved for urban expansion' (Cunha, 2002). It also includes administrative centres of municipios and districts, including suburban zones (United Nations, 2019a). It is very important to mention here that the definition of urban keeps on changing over time with new rounds of census in many countries. The Population Division of United Nations does a very arduous task of adjusting the urban estimates and urban projections as per the new definition. It also provides details of definition used and urban estimates and projections in its regular publication called, 'World Urbanization Prospects'.

In India, during British rule, an urban area was defined as including 'every municipality of whatever size, every cantonment, all civil lines not included in municipal limits, and every other collection of permanently inhabited houses of an urban character with a population of not less than 5000 persons, though not under municipal government' (Census of India, 1911b). This definition of urban remained almost unchanged up to the 1951 census, and left the scope for State Census Superintendents to apply their judgements in declaring the settlements as urban (Bose, 1964). The 1961 census was, thus, a turning point in the evolution of the definition of urban. According to the 1961 census, urban centres included all statutory towns (also known as municipal towns) like Municipal Corporation, Municipality, Notified Area Committee, Cantonment Board, etc. In addition to this, for other settlements not included under any municipal government, more stringent criteria were introduced for a place to qualify as urban (i.e. census town). Three criteria related to population size (5000 and more), population density (at least 400 persons per square kilometre) and workforce structure (at least seventy-five per cent of workers engaged in the non-agricultural sector) were introduced to identify census towns or non-municipal towns for the first time. Thus, from 1961 onwards, the differentiation of space into rural and urban categories along the continuum of 'agriculture-industry' spectrum has become stricter. This definition of urban was adopted as such in the 1971 census. However, in the 1981 census, although urban definition remained the same, it introduced few changes in the measurement of non-agricultural activities. For example, for the first time the 1981 census considered male workers only excluding female workers in the measurement of 75 per cent workers in the non-agricultural activities which is tantamount to introducing gender bias in the definition of urban[3] (Bhagat, 2002; Census of India, 1981a, 1981b). The 1981 census also excluded livestock, forestry, fishing, hunting and plantations, orchards, etc., from the category of non-agricultural activities, whereas these activities were part of non-agricultural activities in the 1961 and 1971 censuses (Census of India, 1981a: 1). Several of these activities promote ancillary industries related to timber, furniture, plywood, tea, coffee, rubber and the like and also boost the trade and business in these products. Many settlements engaged in these types of activities, in fact, satisfied to be as urban (Ramachandran, 1995: 84). Thus, treating many of these categories of workers outside the scope of non-agricultural workforce made urban definition also 'industrially biased' in defining India's urbanisation.

Further the concept of 'work' was adopted for the first time in defining and enumerating workers since the 1961 census compared to the concept of livelihood in earlier censuses. In the 1971 census, more specifically mentioned and adopted the 'main activity' as the basis of work and enumeration of workers. As such, workers belonging to the marginal economic activities/subsidiary work was not included. Realising this deficiency, in the 1981 census along with identification of main workers (those who worked for more than six months during the last one year), also enumerated marginal workers (those who worked for less than six months during the last one year). Against this backdrop of the new classification of workers into main and marginal workers, census took only main workers in the estimation of 75 per cent male workforce in the non-agricultural activities (Office of the Registrar General and Census Commissioner, India (n.d.)). This made the definition even more stringent, and also deepened the gender bias in the definition of urban as the work participation of women in marginal worker's category has been much higher compared to males (Bhagat & Das, 2013: 44). Thus, exclusion of female workforce in the determination of non-agricultural workforce seems unreasonable, however for the sake of comparability over the censuses one should be cautious of recommending any changes in the definition. In addition, the discretion of the Superintendent/Director of Census Operations in declaring a place with distinct urban characteristics as urban in consultation with state governments and the Census Commissioner of India has been an important feature of the censuses since British times (see Appendix I). How the definitional criteria are applied and discretion exercised has also subjected the census for a constructive criticism, notwithstanding Census of India doing a meticulous job (Kundu, 2011).

The urban definition followed in the 2011 census is as follows:

a. Statutory Towns: All administrative units that have been defined by statute as urban like Municipal Corporation, Municipality, Cantonment Board, Notified Town Area Committee, Town Panchayat, Nagar Palika, etc., are known as Statutory Towns.
b. Census Towns: Administrative units satisfying the following three criteria simultaneously are treated as Census Towns:
    i. A minimum population of 5000 persons;
    ii. 75 per cent and above of the male main working population being engaged in non-agricultural pursuits; and
    iii. A density of population of at least 400 persons per sq. km. (1000 per sq. mile).

The categories of urban centres—statutory towns and census towns, have been a significant feature of India's urban definition. There exists a municipal diversity comprising mainly of corporations, Boards,[4] councils, committees, etc., in the municipal governance in India. For each of them, the power, authority and autonomy are assigned by the state governments in their municipal Acts as urban development is a state subject.

It is important to reiterate that the process of identification of census towns based on demographic and work force criteria has been a dynamic process. The stipulated criteria are applied to all places except statutory towns, and it is necessary to satisfy all three criteria namely, population size, density and proportion of non-agricultural workforce. In subsequent census if census towns do not qualify for any of the stipulated demographic and workforce criteria, they will be declassified as villages and treated as rural. Thus, inclusion and exclusion from the category of census towns has been a dynamic process adopted by the Indian census.

As mentioned already, rural–urban frame in India is prepared in advance prior to the actual enumeration of the census. For example, in the census 2011, identification of places that qualify as census towns, was based on data pertaining to size, density and workforce structure from the 2001 census. All villages with a population of 4000 and above (with an assumption that during the intervening decade, the population size would cross the 5000 mark), a density of 400 persons per sq. km. and at least 75 per cent of male main working population engaged in non-agricultural activity in 2001 census was considered as urban.

In India, in situ urbanisation has been an important mode of urbanisation throughout its history. Towns have sprung up as a result of sudden investments by government agencies in certain pockets. Emergence of such urban centres is also facilitated by the administrative mechanism of transplanting district or tehsil headquarters or some other municipal apparatus. In the past, the ruling class established several towns and cities under political and strategic concerns (Census of India, 1961), many of which have survived till this date. The three megacities namely Kolkata, Mumbai and Chennai (formerly Calcutta, Bombay and Madras, respectively) were established as port cities during British rule to facilitate trade and commerce. Building urban centres has been strategic not only for political reasons but for economic strength as well. It is very pertinent to quote Mr. Baille, the British census official, who had said, 'In India, towns attract the trade and not the trade the towns' (Census of India, 1911a).

As a town grows in size over time, urbanisation spreads to the adjoining areas as well. This process is known as suburbanisation. Remarkably, this process is not captured by the existing definition of 'urban' as there is no transitional category in the urban definition. Attention was drawn to the inadequacy of the bilateral classification of places into rural and urban by the Census Superintendent of Punjab way back in 1961. It was argued that the definition of urban was not satisfactory, and it led to underestimation of growth in urban population. It is worthwhile to mention that defining settlements whether urban or rural is not enough and we need to draw a boundary of an urban area to designate it either as a town, city, or an urban agglomeration (UA). In India, an urban centre with a population of more than 100 thousand is generally called a city. The boundaries of a town or a city could be defined either administratively, i.e. municipal boundaries or geographically based on the concept of UA. Indian census defines UA as follows:

An urban agglomeration is a continuous urban spread constituting a town and its adjoining outgrowths (OGs), or two or more physically contiguous towns together with or without outgrowths of such towns. An Urban Agglomeration must consist of at least a statutory town and its total population (i.e. all the constituents put together) should not be less than 20,000 as per the 2001 census. In varying local conditions, there were similar other combinations which have been treated as urban agglomerations satisfying the basic condition of contiguity. There were 474 UAs according to the 2011 census. Examples of UAs are Greater Mumbai UA, Delhi UA, Pune UA, etc. (Census of India, 2011). When concept UA is applied the total number

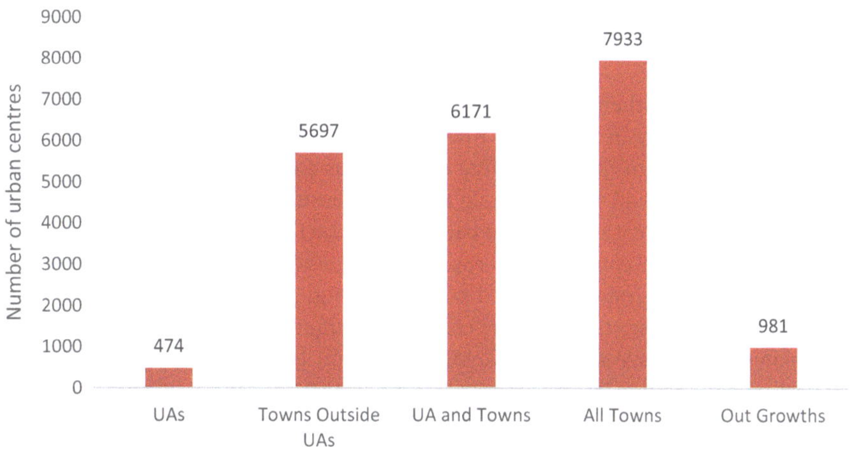


**Fig. 2.1** Number of urban centres based on concepts of UA and towns, India, 2011. Census of India, 2011, A-04: Towns and urban agglomerations classified by population size class with variation between 1901 and 2011, Class I to Class VI; https://censusindia.gov.in/census.website/data/census-tables, accessed on 12 May 2023; Also see UN Statistics (2019). An outgrowth is not counted as a separate town but is an appendage to a statutory town. In the 2011 census the provisional number of towns was 7935, the final number is revised to 7033. Many of these towns are part of UAs and the rest are independent towns

of UAs and Towns were 6171 including 474 UAs in 2011 in contrast to 7933 towns when the concept of town is only used (see Fig. 2.1).

It is clear from the above definition that UA consists of one main urban centre and additional urban centres contiguous to each other including outgrowths. According to the Census of India, an Out Growth (OG) is a viable unit such as a village or a hamlet or an enumeration block made up of such village or hamlet and clearly identifiable in terms of its boundaries and location. Some of the examples are railway colony, university campus, port area, military camp, etc., which have come up near a statutory town outside its statutory limits but within the revenue limits of a village or villages contiguous to the town. While determining the outgrowth of a town, it has been ensured that it possesses the urban features in terms of infrastructure and amenities such as *pucca* roads, electricity, taps, drainage system for disposal of wastewater, educational institutions, post offices, medical facilities, banks, etc., and is physically contiguous with the core town of the UA. Each such town together with its outgrowth(s) is treated as an integrated urban area and is designated as an 'urban agglomeration' (Census of India, 2011). Despite the fact that outgrowths are organically part of a city or town, they are not governed by municipal administration. There were 981 outgrowths according to the 2011 census which formed part of UAs (see Fig. 2.1).

## 2.3 Census Towns or Non-Municipal Towns

In the definition of urban two types of towns are identified by the Census of India namely—(i) statutory or municipal towns and (ii) census or non-municipal towns. More than one-third of the towns were classified as census towns in 1991 and over one-fourth in 2001 at all-India level. In the 2011 census, the share of census or non-municipal towns increased to almost half (49 per cent) of the all-urban centres. This shows the number of census towns rose phenomenally and their share in the total urban centres in the country reached an all-time high (see Fig. 2.2) Out of the net addition of 2530 census towns (new census towns minus declassified town) in 2011, three states namely West Bengal, Kerala and Tamil Nadu alone contributed to 45 per cent of net addition of census towns during 2001 to 2011. It shows the nature of urban transition in India is concentrated and characterised by the resurgence of in situ urbanisation, i.e. transformation of rural settlements into urban resulting from growing population size, rising density, and increasing pace of economic diversification. This process could happen with or without migration known as in situ *urbanisation.* Although the census or non-municipal towns have predominantly non-agricultural base, they have not been accorded municipal status by the respective state governments. In the absence of municipal status, these towns continue to be predominantly governed by *Gram Panchayats* (Bhagat, 2005).

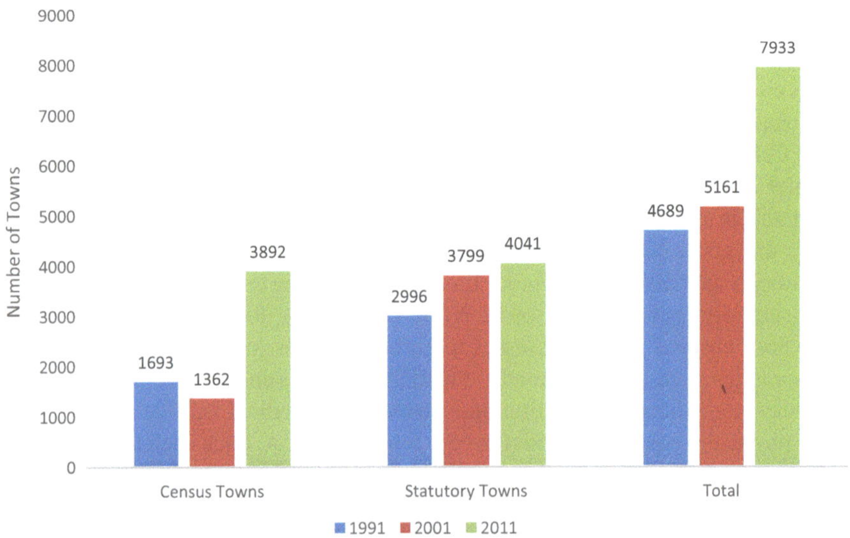


**Fig. 2.2** Number of census and statutory towns, India, 1991 to 2011. Census of India, 2011, A-04: Towns and urban agglomerations classified by population size class with variation between 1901 and 2011, Class I to Class VI; A-04: Appendix 1—New towns added in 2011 and towns of 2001 declassified in 2011 (PC11_PCA_MISC08). https://censusindia.gov.in/census.website/data/census-tables, accessed on 12 May 2023 (*Note* Census was not conducted in Jammu and Kashmir, data for 1991 exclude Jammu and Kashmir)

The state governments have been sluggish in granting municipal status to either census towns or other potential settlements. As per the census figures, at the aggregate level, the decade 2001–2011 witnessed an addition of only 242 municipal or statutory towns across the country. However, the extension of boundaries of the existing municipal urban centres either through the merger of existing urban areas or the extension of boundaries into the rural area has been a regular feature in the organisation of urban settlements in India. Such practices may hugely affect the population size and growth at the town or city levels.

About one-third of the census towns are part of UAs, and the rest of them exist as an independent town. Further, the distribution of new census towns in the 2011 census shows a dispersed pattern of in situ urbanisation. A majority of the census towns (63 per cent) are located away (10 kms or more) from the influence of Class I towns with a population of 100 thousand persons or more (Pradhan, 2013). Some of the leading researchers term such urban processes as a form of *subaltern urbanisation,* i.e. smaller urban settlements existing outside the metropolitan shadow connoting a dispersed pattern of urbanisation. The agency of small urban centres is the core of this concept challenging the metropolitan hegemony in the process of urbanisation (Mukhopadhyay et al., 2020). Paradoxically they also point out a hardly measurable difference in the service provision between the statutory and the census towns when the size of the settlement and effectiveness of governance are controlled (Mukhopadhyay et al., 2019). This justifies the case why state governments are reluctant to grant municipal status to census towns. It is also interesting to note that the nature of rural to urban transition is further rooted in primordial identities and ethos based on caste and religious affiliations persisting particularly in small and medium towns (Vithayathil et al., 2019).

It is also important to emphasise that the metropolitan influence remains significant in shaping India's urbanisation, and researchers have tried to explain the emergence of large number of census towns using the concept of planetary urbanisation developed by Brenner and Schmid (2015). Planetary urbanisation is unleashed by global conditions manifesting in the form of three interlinked processes of concentrated, extended and differential urbanisation. Using this frame, researchers recognise the importance of large urban agglomerations in shaping the extended and differential urbanisation manifesting in the form of census towns (Jain & Korzhenevych, 2022).

While the definition of urban area and rural–urban classification is being debated, it is not generally emphasised that India's urban definition is dynamic, and census or non-municipal towns may also lose their status if they are not able to retain the prescribed size, density and workforce structure. The towns that lose their status in the subsequent censuses are named *declassified towns* by the Census of India. In the 2011 census, the number of declassified towns was 68 which comprised about 0.5 million population in the preceding census. It means that 68 towns of the 2001 census lost their urban status in the 2011 census with a simultaneous loss to the total urban population. In the 2001 census, the number of declassified towns was 445 comprising a total population of about 5 million in the preceding census. Thus, while villages are urbanised if they satisfy the urban criteria, towns lose their status if they are not able to maintain their demographic and economic status.

As mentioned earlier, census towns fall under the jurisdiction of Gram Panchayats, and are entitled for funds under different rural development scheme from the central government. The Ministry of Rural Development of the central government has a huge budget per annum to be allocated to various rural development programmes such as rural employment, housing, cooking gas, water supply, roads, electricity supply, schools and health care. In comparison, funds from Ministry of Urban Development for programmes like poverty alleviation and employment generation are negligible. For many state governments, therefore, maintaining the status quo is advantageous so that they can access the largesse distributed by the central government. In contrast to India where getting reclassified as urban may cause places to lose government transfers, in Egypt getting reclassified as urban would trigger additional public investment for higher-level service delivery requirements, including police stations and courthouses (Dijkstra et al., 2020).

The desire to avoid taxation has also been strong among people in resisting municipalities. The municipal areas have higher taxation as compared to non-municipal areas. In addition, people residing in the municipal areas are subjected to a variety of other taxes and charges. Urban local bodies earn revenue from two major sources namely (i) Tax revenue mainly pertains to property taxes; tax on vehicles, animals, trade and callings and professions; theatre tax/show tax; tax on advertisement, etc., and (ii) non-tax revenue such as rents from municipal assets, income from municipal undertakings, user charges, fees, duties and fines and income from municipal investments. In addition, urban local bodies also receive grants-in-aid from governments.

It is also observed that the electricity charges and tariffs are significantly lower and subsidised particularly among rural households dependent on agriculture and water supply is invariably free, and rural households are also covered under Mahatma Gandhi National Rural Employment Guarantee Scheme (MGNRES)—the largest employment guarantee programme in rural areas. Further, Gram Panchayats are not generating as much revenue from their own sources as compared to municipal bodies. Thus, the Gram Panchayats survive primarily on development grants provided by state governments, and sincere efforts are lacking to generate their own resources owing to low capacity and empowerment.

Many settlements under the influence of big cities are still governed by *Gram Panchayats* where the tax, tariff, and charges are lower. Property taxes are either non-existent or negligible and building rules are hardly applicable. On the other hand, the creation of a municipality is usually followed by some measures of planning and building regulations. It is possible to live in proximity to a city and access its economy, but at the same time escape its planning and building rules. This also favours the builders' lobby ubiquitous in most of the large cities. There are also many instances when settlements were taken out of the municipal limits and officially made villages again (Bhagat, 2005). On the other hand, there is a huge delay or little consideration of granting municipal status to the census towns in almost all states of India in spite of the fact that census towns are declared urban by Census of India.

Thus, the political pragmatism of rural–urban classification vitiates the transition from *panchayat* administration to municipal governance in India.

It is ironical that all settlements with municipal status automatically become urban, while the ones defined as urban based on census criteria do not necessarily get municipal status. At the same time, there were several instances of state governments granting municipal status to settlements possessing conditions hardly different from that of a village. One really wonders why the municipal apparatus was foisted on them (Census of India, 1961). In fact, the grant of statutory status to a settlement is the state prerogative, and should depend upon the level of structural transformation of settlement based on agrarian to non-agrarian activities, financial health as well as the perception of the level of urbanisation. In the 2001 census, the state of Tamil Nadu, a relatively advanced state in southern India had seen an upsurge of statutory towns—an almost seven-fold increase while not a single addition was made in the 2011 census (Bhagat, 2005; Mitra & Kumar, 2015). This shows there is no robust policy of granting municipal status, and decisions are ad-hoc and vary from state to state and also over time. In fact, there are no standard criteria of declaring statutory towns whereas the criteria of identifying non-municipal or census towns are well defined by the census. There is a need to evolve standard criteria for declaring statutory towns. The municipal boundaries of towns need to be changed as and when they expand into the adjoining areas. This is likely to tap the increased revenue potential of the extended urban space. But the state governments do not redraw the municipal boundaries at the earliest nor there is a time frame fixed for it. 'At present, it is a time-consuming process as state governments have to notify such changes through the offices of Deputy Commissioners/District Magistrates in order to invite any objection to the changes in the municipal boundary, which would be duly considered by the state government before declaring the change in it' (Chawla, 1995). As a result, it is not always easy to bring several villages adjacent to large cities/towns within the municipal governance.

In addition, instances of a fairly large built-up area coming up adjacent to the municipal limits of statutory towns in the form of railway colonies, university campuses, port areas, military camps, etc., are not uncommon. As these places are mostly located beyond the statutory limits of towns/cities, they continue to be served by *village panchayats*.

However, such places mostly lie outside the statutory limits of the corporations, municipalities, cantonment boards, etc. In most cases, they usually fall within the revenue limits of the village or villages contiguous to the town. Although, many of them may fail to qualify as independent urban units, for all practical purposes they appear as segments of urban landscape. They are organically very much part of city life, and hence treating them as 'rural' is not only unrealistic but also amounts to underestimation of urban processes. According to the Census of India, they are called 'outgrowths' of a city or town.

## 2.4 Transitional Areas

It is very important to mention that in most of the countries, urban is defined and rural is treated as a residual category. Thus, rural–urban classification is dichotomous, in nature, wherein a host of settlements that are in the transitional stage are either left out of the ambit of the definition of urban or municipal administration in case defined as urban. These transitional areas are also known by various names in urban literature namely suburban (outer part of the city), semi-urban, peri-urban, rural–urban fringe, etc. There is also a term known as *Desakota* in Bahasa Indonesia meaning a co-existence of *desa* (village) and *kota* (city) together (McGee, 1991). It is an area of mixed and varying intensity of rural and urban land uses and economic activities which are interlinked and co-penetrating facilitated by transportation and communication systems. Desakota is one of the earliest concepts on peri-urban areas and has helped a striking advancement in subsequent research. Peri-urban areas are more of recent use. It can be 'described as fringe areas of cities or adjoining rural areas, which are intrinsically linked with the city economy, experience constant transformation, and are characterised by a mix of rural and urban activities' (Aizaz, 2019). Census towns that emerge near an urban core are also part of peri-urban growth, but census towns could exist outside the peri-urban areas. Thus, the peri-urban areas comprise both urban and rural settlements contiguous and interlinked to each other surrounding an urban core. On the other hand, the semi-urban areas are in the process of rural–urban transition but are not defined as urban according to its definition. However, it is worthwhile to mention that these near-urban forms are not mutually exclusive, and are marked by different intensity of rural–urban transition.

Of late, *peri-urbanisation* has emerged as a strong force leading to transformation of landscape in the adjoining rural areas. This process of transformation is more conspicuous near mega cities and state capitals. However, the present classification of rural and urban fails to capture these transitional areas of rural–urban space. Although, as early as 1961, Census of India had recommended 'a three-fold classification of settlements namely, urban, semi-urban and rural' (Census of India, 1961), no attempt was made thereafter to define semi-urban, and hence, the two-fold classification continues till date.

The Constitution of India was amended in 1992. This is known as the seventy-fourth amendment to the constitution, and this amendment envisages three types of urban local bodies namely *Nagar Panchayat,* Municipal Council and Municipal Corporation. A Nagar Panchayat (or by whatever name it is called) shall look after the governance of transitional areas in the close vicinity of cities or towns. Likewise, a Municipal Council shall provide governance in smaller urban areas while a Municipal Corporation is designed for larger urban areas. With regard to the definition of 'transitional areas', 'smaller urban areas' or 'larger urban areas' the amendment states that the Governor of the respective state 'may having regard to the population of the area, the density of population therein, the revenue generated for local administration, the percentage of employment in non-agricultural activities, and the economic importance of other factors as he/she deems fit, specify by public notification for this purpose' (Govt. of India, 1992).[5] Remarkably, the population size for different local bodies has not been specified, and the same has been left to the discretion of the respective state governments. In the absence of information on the transitional areas from the Census of India, the state governments apply their own judgement. For example, The Maharashtra Municipal Councils, Nagar Panchayats, and Nagar Panchayats Act provide that no area shall be specified as 'transitional area' unless it has a population of not less than 10,000, and not more than 25,000, and is within a distance of 25 km from the territorial limits of any Municipal Corporation or 'A' Class Council, and at least one-fourth of employment is in non-agricultural activities. If any transitional area lies more than twenty kilometres away from the territorial limits of any Municipal Corporation or an 'A' Class Council, but the percentage of employment in non-agricultural activities in such area is not less than fifty per cent, it shall be accorded the status of Nagar Panchayat (Govt. of Maharashtra, 2016). The size of population for Nagar Panchayat (or Town Panchayat) differs from one state to another, and percentage of non-agricultural employment is often not specified.

It will be more than appropriate, if the census attempts to identify not only rural and urban areas but also transitional areas. This requires necessary modifications in the definition of rural and urban by Census of India with a separate category of 'semi-urban' in view of the requirements of decentralised governance as envisaged in the 74th Amendment to the Constitution. Census could also provide certain relevant information such as property taxes, fees, service charges, etc., paid to the local government in its Household Tables published extensively since the 1981 census. Economic environment is undergoing a rapid transformation, and, therefore, such data will play a crucial role in monitoring the economic health of the local governments. It will also be helpful for the state governments in assessing the viability of urban centres for municipal governance.

It goes without saying that a precise definition of transitional urban areas is essential for the state governments in achieving the goal envisaged in 74th Amendment to the constitution. In addition, a mechanism for regular assessment of revenue potential of non-municipal urban centres as well as outgrowth of cities and towns must be evolved. This will facilitate grant of civic status to urban centres and revision of existing municipal boundaries of statutory towns in an objective manner. Regular assessment of revenue potential is also necessary for finalising funds for devolution to the local governments. In practice, the mechanism of granting a municipal status to a settlement diverges from state to state, and the state governments also exercise their discretion in including an area in any category of municipal status. The state governments can also include Gram Panchayats consisting of many villages or a village or a part of a village in the municipalities. As such there is nothing to limit the power of the state governments in constituting municipal areas (Chawla, 1995). Thus, a proper rationale for municipal organisation of space corresponding to rural–urban classification is lacking and needs to be rectified after each census by the respective state governments.

## 2.5 Distinguishing Between Cities and Towns

The Census of India provides a six-fold classification of urban centres[6] which is categorised into varying sizes distinguishing initially between cities and towns. In the Indian context, cities are defined as urban settlements having 100,000 or more population, while all other urban settlements with a population size of less than 100,000 persons are designated as towns. Towns are further grouped under three categories viz. small towns, medium towns and large towns. Small towns refer to urban settlements having less than 20,000 population. Likewise, medium and large towns have a population size of 20,000–49,999 persons, and 50,000–99,999 persons, respectively (Census of India, 1991). There are two ways of drawing the boundaries of cities and towns. One is the administrative or municipal boundaries which are drawn by the state government that designate cities and towns as Municipal Corporation, Municipal Council/Town Council/Municipality, Nagar Panchayat/Town Panchayat, etc.[7] The municipal act specifies the designation and delineation of various municipal entities. Another is the boundaries drawn based on Urban Agglomeration (UA) as defined and determined by the Census of India. Internationally, UA is considered as the standard definition for determining the size of a city or an urban centre. It is, therefore, very important to keep in mind the difference in determining the boundaries and size of an urban centre while comparing with others or calculating the growth rate of the same entity over time. For example, the population size of Mumbai (Municipal Corporation of Greater Mumbai known as MCGM) was 12.4 million while that of Mumbai (Urban Agglomeration) was 18.4 million as per the 2011 census. On the other hand, the Mumbai Metropolitan Region (MMR) created by an Act of the Government of Maharashtra for the planning purpose comprises of 23.5 million population as per the 2011 Census (https://mmrda.maharashtra.gov.in/about-mmr, accessed on 16 August 2023). A similar multi-level categorisation is also applicable to other cities as well. Therefore, one can see that cities exist in multiple entities and one should compare the same entity. It is also very important to mention that cities exist as an unbounded category in the form of agglomeration, and their delineation is undertaken by administrative, municipal and planning processes.

There are a large number of statutory towns that fall under the category of small and medium towns (below 50,000) and are governed by municipal councils/town councils and Nagar Panchayats. Understandably, most of the census towns belong to small size category (i.e. below 20,000), and their population base is not viable for municipal governance. A town group approach, similar to the one adopted in the 1961 census, is likely to serve as a useful framework for urban governance in the present context as well. It may be recalled that the 1961 census had, for the first time, provided statistics both separately as well as collectively for towns located closer to one another over space, and despite several ULBs, they behaved as one single unit in daily activities. The underlying idea was that grouping of urban centres in that manner would prove useful not only for governance but also in development planning of the entire region. This concept was renamed as urban agglomeration (UA) in the 1971 census (Census of India, 1971), which continued in the later censuses also. However, it may be noted that the concept of UA has not been utilised in urban governance. In 1971, the concept of Standard Urban Areas (SUA) was also introduced which had the potential to meet the emerging challenges of viability.[8] Delimited with respect to urban centres with a population of 50,000 or more, SUA was the projected urbanised areas of the town/city over the next 20 years which included not only the adjoining towns but also villages based on the select parameters (Census of India, 1981b). They are as follows:

i. Predominant urban land use,
ii. Intensive interaction with urban centres as reflected in commutation for the purpose of work and secondary education facilities; extension of city bus service; sale of commodities like milk, dairy products, vegetables (other than those transported by rail or truck) and purchase of food grains, clothes, general provisions, etc., by the consumers directly.
iii. Anticipated urban growth as a result of locational decisions relating to industry, market, transport and communication, administrative and service functions.
iv. Existence of big villages with a large proportion of the workforce engaged in non-agricultural industrial categories.

Although the concept of SUA was dropped in subsequent censuses, the idea could be used in reorganising urban space in light of new requirements of municipal governance. This is also likely to minimise revenue loss and evasion of taxes on the account of being rural within the area of influence of an urban centre. For metro cities instead of city centric governance, metropolitan governance consisting of contiguous urban agglomerations and adjoining rural areas could be another alternative (Biswas, 2020). The local bodies don't necessarily need to be either completely rural or completely urban but an amalgam of the two could also be a better choice, keeping in view the rural and urban relationship and interdependence. However, the impact of the spatial reorganisation of municipal governance will depend much upon the extent of autonomy, empowerment and capacity building of the municipal bodies.

## 2.6 Concluding Remarks

There is no universal definition of urban at the global level. It is neither desirable nor essential to have a universal definition as urban represents diversity of economic and social conditions of nations passing through different stages of development. However, a global definition of cities, towns and rural areas has been proposed recently endorsed by UN Statistical Commission. This definition takes into account only size and density in defining cities (50,000 inhabitants or more with a grid density of 1500 persons per $km^2$ and more), towns (5000 inhabitants with grid density of not less than 300 persons per $km^2$) and rural areas which consist of the low-density grid with less than 300 per $km^2$ (Dijkstra et al., 2020). However, the grid-based approach of population size and density in defining urban ignores the administrative and political framework of local governance as well as the historical evolution of settlements designated as cities and towns.

This chapter has examined rural–urban classification in India in the context of renewed emphasis on local self-government as envisaged in the 74th Constitutional Amendment. The said amendment visualises three types of urban governance corresponding separately to larger, smaller and transitional urban areas, respectively. Interestingly, however, no criteria could be specified because municipal governance is a state subject in the Indian federal system, and states apply their own criteria in designating urban areas for municipal governance. The Census of India also does not provide information on transitional areas which are semi-urban in nature. The constitutional amendment mandates that the transitional areas shall be governed by Nagar Panchayats.

A number of census towns also exist outside the ambit of municipal governance. It should be realised that what constitutes urban in the country is not necessarily municipal, while all municipal areas are defined as urban. Non-agricultural activities form the base of economy of census or non-municipal towns. Governed by rural local bodies (i.e. *Gram Panchayats*), they are hardly empowered to generate their resources for governance. Instead, they continue to be rewarded with the funds from various rural development programmes. This technically means diversion of funds meant for the rural areas to the areas that are not truly rural in character. At the same time this also means that the revenue potential of these non-municipal towns is not being realised in want of municipal governance. Further, any revision of municipal boundaries conforming to urban transformation is a very time consuming process in the country. This unusual delay results in the loss of revenue, as the urbanised areas are not brought within the municipal limit at the right time.

The viability of small towns is a matter of concern in view of their important role in municipal governance. The concept of SUA or UA could be useful in achieving their viability. On the other hand, for metro cities, metropolitan governance could be a viable alternative to address many issues of sustainability, convergence of planning and programme implementation. However, the commitment and sincerity with which the state governments empower ULBs, both politically and financially, shall determine the success of spatial rationalisation. The erstwhile Planning Commission was aware about it when it noted that the 'urban governance today is characterised by fragmentation of responsibility, incomplete devolution of functions and funds to the elected bodies and ULBs (urban local bodies), unwillingness to progress towards municipal autonomy, adherence to outmoded methods of property tax and reluctance to levy user charges' (Planning Commission, 2002).

An inclusive and sustainable urbanisation based on the principles of 74th Amendment to the Constitution and SDG Goal 11 (Sustainable Cities and Communities) should be the basis of what constitutes urban and the objective of rural–urban classification for governance, planning and the well-being of the people.

**Acknowledgements** This is thoroughly revised version of the article of the first author entitled, 'Rural-urban classification and municipal governance in India', *Singapore Journal of Tropical Geography*, 2005, 26 (1), 61–73. Re-used here as per the rights vested with the Author.

# Appendix I

## *Changing Criteria of Defining Places as Urban*

| | | |
|---|---|---|
| 2011 | A. All places with a municipality, corporation, cantonment board or notified town area committee, etc. (known as Statutory Town) | Male main workers only |
| | B. All other places which satisfied the following criteria (known as Census Town): | |
| | A minimum population of 5000; | |
| | At least 75 per cent of the male main workers engaged in non-agricultural pursuits; and | |
| | A density of population of at least 400 per sq. km | |
| 2001 | Municipal Corporation | It is clarified that the measurement of male workers in non-agricultural activities is based on main workers only which means it excludes marginal workers |
| | Municipal Area | |
| | Town Committee | |
| | Notified Area Committee | |
| | Cantonment Board | |
| | All other Places satisfied the following criteria | |
| | (i) A minimum population of 5000 | |
| | (ii) At least 75% of the male working population engaged in non-agricultural pursuits | |
| | (iii) A population of at least 400 per Sq.km (or 1000 per Sq.mile) | |
| 1991 | Municipal Corporation | 1981 Census introduced two changes: (i) In working out the proportion of male workers in non-agricultural activities such as livestock, forestry, fishing, hunting and plantations, orchards and allied activities have been treated as agricultural activities. In 1961 and 1971 censuses these activities had been categorised as non-agricultural (ii) Only male workers included in working out non-agricultural activities |
| | Municipal Area | |
| | Town Committee | |
| | Notified town Area Committee | |
| | Cantonment Board | |
| | All other places satisfying the following criteria: | |
| | (a) Density of population of at least 400 per sq.km | |
| | i.e. (1000 persons per sq. mile) | |
| | (b) A minimum population of 5000 | |
| | (c) 75% of the male working population engaged in non-agricultural (allied) activities | |
| 1981 | Municipal Corporation | |
| | Municipal Area | |
| | Town Committee | |
| | Notified town Area Committee | |

(continued)

(continued)

| | | |
|---|---|---|
| | Cantonment Board | |
| | All other places satisfying the following criteria: | |
| | (a) Density of population of at least 400 per sq.km | |
| | i.e. (1000 persons per sq. mile) | |
| | (b) A minimum population of 5000 | |
| | (c) 75% of the male working population engaged in non-agricultural (allied) activities | |
| 1971 | Municipal Corporation | |
| | Municipal Area | |
| | Town Committee | |
| | Notified town Area Committee | |
| | Cantonment Board | |
| | Selected places with | |
| | (a) Density of population of not less than 400 persons per sq. km | |
| | (b) A population of 5000 | |
| | (c) three-fourth of working population should be working outside agriculture | |
| | (d) or any other place, which according to the Director of Census Operations possesses pronounced urban characteristics and amenities | |
| 1961 | Municipal Corporation | |
| | Municipal Area | |
| | Town Committee | |
| | Notified town Area Committee | |
| | Cantonment Board | |
| | Selected places with | |
| | (a) Density of population of not less than 1000 persons per sq. mile | |
| | (b) A population of 5000 | |
| | (c) three- fourth of working population should be working outside agriculture | |
| | (d) or any other place, which according to the Superintendent of State possesses pronounced urban characteristics and amenities | |
| 1951 | Municipal Corporation | |
| | Municipal Area | |
| | Town Area Committee | |
| | All civil lines not included in Municipal Limits | |
| | Cantonment | |

(continued)

(continued)

| | | |
|---|---|---|
| | Test prescribed for distinguishing towns from villages in different states were based on ideas common to all states, but they were not identical nor had they been applied with meticulous uniformity. In the case of Princely States, the definition of town was applied a little indiscriminately | |
| 1941 | Municipal Corporation | |
| | Municipal Area | |
| | All civil lines not included in Municipal Limits | |
| | Cantonment | |
| | Every continuous collection of houses inhabited by not less than 5000 persons, which the Provincial Superintendent may decide to treat as a town | |
| 1931 | Municipal Corporation | |
| | Municipal Area | |
| | All civil lines not included in Municipal Limits | |
| | Cantonment | |
| | Every other continuous collection of houses inhabited by not less than 5000 persons, which the Provincial Superintendent may decide to treat as urban. In making this decision consideration was given to | |
| | (a) the character of the population | |
| | (b) relative density of the dwellings | |
| | (c) importance in trade, historic associations and to avoid treating as towns overgrown villages without urban characteristics | |
| 1921 | Municipal Corporation | |
| | Municipal Area | |
| | All civil lines not included in Municipal Limits | |
| | Cantonment | |
| | Every other continuous collection of houses inhabited by not less than 5000 persons, which the Provincial Supcrintendent may decide to treat as town for census purposes. In Indian states where there were no municipalities, this definition was required to be extensively applied. Note: In dealing with questions arising under head (6) the Provincial Superintendent will have regard to the character of the population, the relative density of the dwellings, the importance of the place as a centre of trade and its historic associations and will bear in mind that it is undesirable to treat as towns overgrown villages, which have no urban characteristics | |
| 1911 | Municipal Corporation | |
| | Municipal Area | |
| | All civil lines not included in Municipal Limits | |

(continued)

(continued)

| | | |
|---|---|---|
| | Cantonment | |
| | Every other continuous collection of houses inhabited by not less than 5000 persons, which the Provincial Superintendent chiefly may decide to treat as town for census purposes. Note: However, a few places, chiefly in the native states which did not satisfy the requirements were treated as towns for special purposes. The Provincial Superintendents were, however, instructed when considering the question of treating places as towns on the basis of their population, to take care to, exclude such as are merely overgrown villages and have no urban features. It is true that the discretion thus allowed has occasionally led to a certain want of uniformity | |
| 1901 | Municipal Corporation | |
| | Municipal Area | |
| | All civil lines not included in Municipal Limits | |
| | Cantonment | |
| | Every other continuous collection of houses permanently inhabited by not less than 5000 persons, which the | |
| | Provincial Superintendent may decide to treat as town for census purposes | |
| 1881 | Municipal Corporation | |
| | Municipal Area of any place brought under some similar regulation for police or sanitary purposes | |
| | Cantonment | |
| | (a) Every place where the proportion of the trading and industrial population to the total is equal or greater than that of the agricultural population | |
| | (b) Every other continuous group of houses inhabited by not less than 5000 people | |

*Source* Census of India (2001a, 2001b), Census of India (2011), and Office of the Registrar General and Census Commissioner, India (n.d.).

## Notes

1. Some countries like the Philippines had set up a special committee to study the criteria of rural-urban classification. About the validity of such criteria, the report mentions that users of the urban definition would be the best judge of its efficiency and applicability (National Statistics Office, Republic of the Philippines 1996).
2. The hukou system in China was introduced in 1958—a hukou book was assigned to each household to classify the resident's hukou by their place of origin. The hukou system categorises each Chinese citizen as either an agricultural (rural) hukou holder or a non-agricultural (urban) hukou holder. Each type of hukou enjoys different types of benefits and specific rights. It is an inherited

status and privilege. The households with rural hukou are distributed arable land for their livelihood while urban hukou holders have access to government jobs, subsidised housing, education and health care. https://www.china-briefing.com/news/chinas-hukou-system/, accessed on 21 May 2023.

3. Although local government belongs to the state list, the delimitation of cantonment areas and formation of local government in it falls under the union list as per provisions of the Constitution. Cantonment harbours both military and civil population. The cantonment administration is under the control of Directorate General Defence Estate (DGDE), Ministry of Defence, Govt. of India. Cantonment areas are better in terms of green space and some of the amenities. There are 61 Cantonments in the country (https://www.dgde.gov.in/content/cantonments, accessed on 20th August, 2023).

   It is also to be noted that the nomenclature of municipal governance has a variety of names. The power and responsibilities are decided by the respective state government.

| | | | |
|---|---|---|---|
| C.B | Cantonment Board/Cantonment | N.M | Non-Municipal Town |
| I.T.S | Industrial Township | T.C | Town Committee/Town Area Committee |
| E.O | Estate Office | N.T | Notified Town |
| G.P | Gram Panchayat | N.T.A | Notified Town Area |
| I.N.A | Industrial Notified Area | S.T.C | Small Town Committee |
| C.M.C | City Municipal Council | N.P | Nagar Panchayat |
| M | Municipality | T.M.C | Town Municipal Council |
| M.B | Municipal Board | T.P | Town Panchayat |
| M.C | Municipal Committee | T.S | Township |
| M.Cl | Municipal Council | C.T | Census Town |
| M.Corp | Municipal Corporation/Corporation | U.A | Urban Agglomeration |
| N.A | Notified Area | O.G | Outgrowth |
| N.A.C | Notified Area Committee/Notified Council | | |

4. There is no reason given by Census of India why only male workforce is taken into account in the definition of urban. The reason seems to be due to underestimation of women workforce in the censuses. The female workers have been highly underestimated in some states like Punjab and Haryana.
5. It was provided in the Act that a municipality may not be constituted in such urban area or part thereof as the Governor may having regard to the size of the area and the municipal services being provided or proposed to be provided by an industrial estate in that area and such other factors as he deems fit, by public notification, specify to be an industrial township.
6. Census of India provides six-fold classification of urban centres based on population size as follows:

   Class 1 (more than 1,00,000).
   Class II (50,000 to 99,999).
   Class III (20,000 to 49,999).
   Class IV (10,000 to 19,999).
7. Class V (5000 to 9,999).

   Class VI (less than 5,000).

   In India, urban local bodies are not hierarchical. However, the Municipal Corporation, as an institution, enjoys a greater measure of autonomy than other forms of local government. It enjoys the power of dealing directly with the state government whereas the municipalities have no direct access to the state government and are answerable to the District Collectors and Divisional Commissioners (UNESCAP 1999).

8. Amalgamation of municipalities and communes has also been happening in the developed countries as well (Goldsmith 1993).

## References


Aizaz, R. (2019). *India's Peri-Urban Regions: The Need for Policy and the Challenges of Governance* (ORF Issue Brief, No. 285). Observer Research Foundation.

Bhagat, R. B. (2002). Challenges of Rural-Urban Classification for Decentralised Governance. *Economic and Political Weekly, 37*(25), 2413–2416.

Bhagat, R. B. (2005). Rural-Urban Classification and Municipal Governance in India. *Singapore Journal of Tropical Geography, 26*(1), 61–73.

Bhagat, R.B. (2016). Changing Pattern of Internal Migration in India. In C.Z. Guilmoto & G.W. Jones (Eds.), *Contemporary Demographic Transformations in China, India and Indonesia.* Springer. https://doi.org/10.1007/978-3-319-24783-0_15.

Bhagat, R.B., & Das, K.C. (2013). *Levels, Trends and Structure of Workforce in India.* Gyan Publishing House.

Biswas, A. (2020). Establishing Metropolitan Governance and Local Governance Simultaneously: Lesson from India's 74th Constitutional Amendment Act. *Journal of Urban Management, 9*(2020), 316–330.

Bose, A. (1964). A Note on the Definition of 'Town' in the Indian Census: 1901–1961. *The Indian Economic and Social History Review, 1*(3), 84–94.

Brenner, N., & Schmid, C. (2015). Towards a new epistemology of urban? *CITY, 19*(2&3), 151–182.

Census Bureau. (2022). Urban Criteria for the 2020 Census: Final Criteria. https://www.federalregister.gov/documents/2022/03/24/2022-06180/urban-area-criteria-for-the-2020-census-final-criteria. Accessed 25 May 2023.

Census of India. (1911a). Vol. XV, *United Province of Agra and Oudh,* Part I, Report by E.A.H. Blunt, ICS, Superintendent Census Operations (Reprint Usha, New Delhi, 1987).

Census of India. (1911b). Vol. 1, *India,* Part II, Tables by E.A. Gait, Superintendent, Govt. Printing India, Calcutta, 1913 (Usha, New Delhi, 1987).

Census of India. (1961). *Punjab, General Report,* Vol. XIII, Part I-A (i), by R.L. Anand, Superintendent of Census Operations and Enumeration Commissioner.

Census of India. (1971). *General Population Tables*, Part II-A(i), Series I, India, New Delhi: Office of the Registrar General and Census Commissioner, Government of India.

Census of India. (1981a). *General Population and Population of Scheduled Castes and Scheduled Tribes* (Paper 2 of 1984, Series 1, India). Registrar General and Census Commissioner.

Census of India. (1981b). Series 1, India, Part-ii-A (iii), *Standard Urban Areas.* Registrar General and Census Commissioner.

Census of India. (1991). *Emerging Trends of Urbanisation in India.* Occasional Paper No. 1 of 1993, prepared by Jain, M.K., Ghosh, M. and Kim, W.B., Office of the Registrar General, and Census Commissioner.

Census of India. (2001a). *A-02 Statement 4: Progress in Urban Population of States and Union Territories 1901–2001.* Office of the Registrar General and Census Commissioner. https://censusindia.gov.in/census.website/data/census-tables. Accessed 16 November 2024.

Census of India. (2001b). *General Population Tables,* Karnataka (A-1 to A-4) (pp. 31–32). Directorate of Census Operations.

Census of India. (2011). Meta Data, office of the Registrar General and Census Commissioner.

Champion, T., & Hugo, G. (2003, May 1–3). Beyond the Urban-Rural Dichotomy: Towards a New Conceptualisation of Settlement for Demographers. *Paper Presented in PAA 2003 Annual Meeting.*

Chawla, B.S. (1995). The Haryana Municipal Act 1973, (Commentary with Allied Rules and Bye-laws, Amended Up-to-date) Chandigarh, Chawla Publications Pvt. Ltd.

Chen, X., Orom, H., Hay, J.L., Waters, E.A., Schofield, E., Li, Y., Kiviniemi, M.T. (2019). Differences in Rural and Urban Health Information Access and Use. *Journal of Rural Health, 35*(3), 405–417. https://doi.org/10.1111/jrh.12335. Epub 2018 Nov 16. PMID: 30444935; PMCID: PMC6522336.

Congress of the US, Office of Technology Assessment. (1989). Defining Rural Areas: Impact on Health Care Policy and Research Staff Paper. www.wws.princeton.edu/~ota/disk1/1989/8912_n.html

Cunha, J. M. (2002, March 11–15). Processes of Urbanisation and Metropolitanisation in Brazil between 1970 and 2000: Continuity, Ruptures and Methodological Challenges. *Paper presented in the Conference New Forms of Urbanisation: Conceptualising and Measuring Human Settlement in the Twenty First Century Organised by IUSSP Working Group on Urbanisation and held at Rockefeller Foundation Study and Conference Centre.*

Dijkstra, L., Hamilton, E., & Lall, S. (2020, March 10). How Do We Define Cities, Towns, and Rural Areas? World Bank Blog. https://blogs.worldbank.org/sustainablecities/how-do-we-define-cities-towns-and-rural-areas. Accessed 16 August 2023.

Goldsmith, M. (1993). Europeanisation of Local Government. *Urban Studies, 30*(4–5), 683–699.

Govt. of India. (1992). The Constitution Seventy-Fourth Amendment Act of 1992 on Municipalities. Ministry of Urban Development.

Govt. of Maharashtra. (2016). The Maharashtra Municipal Councils, Nagar Panchayats and Nagar Panchayats Act 1965 (As modified up to 11th May 2016), Printed in India by the Manager, Government Central Press, Mumbai and Published by the Director, Government Printing, Stationery and Publications.

Halfacree, K.H. (1996). From Locality to Social Representation: Space, Discourse and Alternative Definition of the Rural. In C. Hamnett (Ed.)m *Social Geography: A Reader.* Arnold.

Huang, G. (1995). National Data for Studying Rural Education. www.ed.gov/databases/ERIC_digest/ed383518.html

Hugo, G., Champion, A., & Lattes, A. (2003). Toward a New Conceptualization of Settlements for Demography. *Population and Development Review, 29*(2), 277–297. http://www.jstor.org/stable/3115228

Jain, M., & Korzhenevych, A. (2022). The Concept of Planetary Urbanization Applied to India's Rural to Urban Transformation. *Habitat International, 129*, Article 102671. https://doi.org/10.1016/j.habitatint.2022.102671 (online version 26 September 2022).

Kundu, A. (2011). Method in Madness: Urban Data from the 2011 Census. *Economic and Political Weekly, 46*(40), 13–16.

McGee, T. G. (1991). Emergence of Desakota Regions in Asia: Expanding a Hypothesis. In N. Ginsburg, B. Koppel, & T.G. McGee (Eds.), *The Extended Metropolis: Settlement Transitions in Asia.* University of Hawaii Press.

Mitra, A., & Kumar, R. (2015). New Patterns of Indian Urbanization; the Emergence of Census Towns. *Environment and Urbanization Asia, 6*(1), 18–27.

Mukhopadhyay, P., Zerah, M., & Denis, E. (2019). Subaltern Urbanisation Revisited. In S. Patel & O. Goyal (Eds.), *India's Contemporary Urban Conundrum* (pp. 28–44). Routledge India, South Asia Edition.

Mukhopadhyay, P., Zerah, M., & Denis, E. (2020). Subaltern Urbanisation: Indian Insights for Urban Theory. *International Journal of Urban and Regional Research, 44*(4), 582–598. https://doi.org/10.1177/0975425315583755

OECD/China Development Research Foundation. (2010). Trends in Urbanisation and Urban Policies in OECD Countries: What Lessons for China? *OECD Publishing.* https://doi.org/10.1787/9789264092259-en

Office of the Registrar General and Census Commissioner, India. (n.d.). Meta Data. https://new.census.gov.in/census.website/node/461. Accessed 17 November 2024.

Pandey, V. N. (2003). Re-Presenting Rural: From Definition to Discourse. *Sociological Bulletin, 52*(1), 32–52.

Planning Commission. (2002). *Tenth Five Year Plan, 2002–2007*, Vol. II, Sectoral Policies and Programmes (p. 613). Govt. of India.

Pradhan, K. (2013). Unacknowledged Urbanization: The New Census Towns of India. *Economic and Political Weekly, 48*(36), 43–51.

Ramachandran, R. (1995). *Urbanization and Urban Systems in India.* Oxford University Press.

Reza, A. (2002). Understanding Urbanisation. *Economic and Political Weekly, 37*(44/45), 4554–4555.

The Indian Institute of Public Opinion (IIPO). (2001). Urban Governance. *Monthly Public Opinion, 46*(12), 14–16.

U.S. Bureau of Census. (2001). Rural-Urban Classification Criteria for 2000 Census. *Federal Register, 66*(60), 17017–17033 (Federal Register on line via GPO Access [wais.access.gpo.gov].

UN Statistics. (2019). Expert Group Meeting on Statistical Methodology Delineating Cities and Rural Areas, A Presentation by NSO. https://unstats.un.org/unsd/demographic-social/meetings/2019/newyork-egm-statmeth/docs/s09-02-IND.pptx. Accessed 18 November 2024.

UNESCAP. (1999). Local Government in Asia and the Pacific: Country Paper-India. http://www.unescap.org/huset/lgstudy

UN-Habitat. (2002, March). Global Campaign on Urban Governance. Concept Paper, United Nations Human Settlements Programme (UN-Habitat) (p. 26). http://www.unhabitat.org/compaign//governance/documents/concept_paper.doc

United Nations. (2001). *World Urbanisation Prospects*: The 1999 Revision, Population Division, Department of Economic and Social Affairs, U.N.

United Nations, Department of Economic and Social Affairs, Population Division. (2019a). *World Urbanization Prospects: The 2018 Revision* (ST/ESA/SER.A/420). United Nations.

United Nations, Department of Economic and Social Affairs, Population Division. (2019b). *World Urbanization Prospects 2018*: Highlights (ST/ESA/SER.A/421).

Vithayathil, T., Singh, G., & Pradhan, K. (2019). Only 'good people', Please: Residential Segregation in Urbanising India. In S. Patel & O. Goyal (Eds.), *India's Contemporary Urban Conundrum* (pp. 45–54). Routledge India, South Asia Edition.

Wirth, L. (1938). Urbanism as a Way of Life. *American Journal of Sociology, 14*, 1–24.

Zhu, Y. (2017). *In situ* Urbanization in China: Processes, Contributing Factors, and Policy Implications. *China Population and Development Studies, 1*, 45–66. https://doi.org/10.1007/BF03500917

Zhu, Y. (2001, August 18–24). The Transformation of Township into Towns and Their Roles in China's Urbanisation: Evidence from Fujian Province. In *Paper Presented in 24th IUSSP General Conference.*

# Chapter 3
# Levels, Trends and Patterns in India's Urbanisation

## 3.1 Introduction

Urbanisation is a socio-economic process that transforms rural into urban settlements, expands built-up area and shifts the population from rural to urban areas. It is associated with transition in occupation mainly based on agriculture to non-agricultural activities, changes in lifestyle, culture and behaviour (United Nations, 2019a). Urbanisation is about how characteristics of settlements are changing, and spaces and places are transformed in consonance with economic transition. Urbanisation has emerged as a global force with 55 per cent population of the world living in urban areas by 2018. It is associated with rising GDP, transformation of agriculture to non-agricultural sector, expansion of infrastructure and changes in the characteristics of population. Urbanisation embodies the forces of agglomeration shaping place and space economy through rising density, reducing distance and overcoming divisions promoting economic growth (World Bank, 2009). It is also a process that manifests in the form of varying sizes of cities and town and their hierarchy and relationship closely connected with regional economic processes of convergence and divergence. Historically, the Industrial Revolution in the Western world has intensified the speed of urbanisation leading to the expansion of infrastructure such as transport and communication and increased rural to urban migration. The agglomeration of population, predominance of non-agricultural activities and better provision of social amenities including health and educational infrastructure emerged as distinguishing features of settlements following industrialisation of agrarian economies (Bhagat, 2005). The similar process has unfolded in several parts of the developing countries with colonial expansion. In the study of urbanisation it is pertinent to know how urban areas are defined because, from the demographic point of view, the level of urbanisation is measured in terms of percentage of population living in urban areas (Davis, 1962). An area is classified as rural and urban depending upon various criteria such as population size, density, occupational composition and civic status. There is no thumb rule to divide rural and urban, and the practice is followed diversely



R. B. Bhagat and M. I. Hassan, *Urbanisation and Urban Policies in India*, Exploring Urban Change in South Asia, https://doi.org/10.1007/978-981-96-3181-0_3

across the countries of the world. The definitional issues are discussed in detail in Chapter 2.

Urbanisation is measured by degree of urbanisation (also known as level of urbanisation) typically expressed as the percentage of population residing in urban areas. We can say a country is urbanising when the degree of urbanisation has been increasing. Urban population growth alone is not a sufficient criterion for urbanisation but urban population growth must be higher than the rural population growth. So demographically speaking when percentage urban is increasing, the percentage rural will decline. This turns urbanisation into a finite process. However, as urbanisation also manifests in number and the size of cities, it signifies increasing concentration of population in bigger cities unless urban policies and programmes attempt to reverse it. Looking from this perspective urbanisation is an infinite process of increasing concentration of population and agglomeration of economic activities. However, the concentration of population and economic activities may decline in large cities while increasing in small urban centres as people migrate from bigger cities to small urban centres. This may occur without the level of urbanisation declining. On the other hand, if there is a reversal of rural to urban migration, as rural areas are better served and have good living and working environment, the percentage of population living in urban areas may decline. Such processes are contrary to the process of urbanisation known as counter-urbanisation or deurbanisation (Champion, 1999; Codrington, 2005). Thus, a country may pass through various stages of urbanisation, deurbanisation and reurbanisation characterising an urbanisation cycle.

It is also important to distinguish between Urban Population Growth Rate and Urbanisation Rate. It is possible to have rapid urban population growth rate without having rapid rate of urbanisation. Urban population growth rate depends upon population change in two points of time while urbanisation rate is the change in the degree or level of urbanisation. Most of the countries of the world have experienced massive change in the degree or level urbanisation during the last century. A merely 13 per cent of the global population was urban in 1900, which increased to 29 per cent in 1950 and crossed the 50 per cent mark (50.1 per cent) in 2009, and reached 55 per cent in 2018 (UN, 2009, 2019a). By mid-century the level of urbanisation is likely to reach 68 per cent—roughly two-thirds of the world population, i.e. 9.77 billion (UN, 2019a). But the pattern of urbanisation is found to be very uneven between the more developed and less developed world. Most of the urban population growth is likely to accrue in Asia and Africa. About seventy eight per cent people of the developed world live in urban areas compared to 50 per cent in the less developed ones in 2018. Most of the parts of the world are highly urbanised except Asia and Africa. In Asia 5 out of 10, and in Africa, 4 out of 10 persons lived in urban areas in 2018. These two continents have similar paths of urbanisation and also have low levels of per capita income (Cohen, 2004; United Nations 2019a). Within Asia, Southern Asia has the lowest level of urbanisation, i.e. 36 per cent compared to 72 per cent in West Asia and 63 per cent in Eastern Asia including China (59 per cent) and Japan (92 per cent). Southern Asia produced only 8 per cent of global GDP with a share of the global urban population of 14 per cent (a ratio of 0.57) (Ellis & Roberts, 2016: 2). In Southern Asia, the level of urbanisation ranging from 38 per cent in Bhutan to 18 per cent in Nepal. The UN estimates show the level of urbanisation in India is 34 per cent compared to 37 per cent in Bangladesh and also 37 per cent in Pakistan (United Nations, 2019b).

As discussed in Chapter 2, there is a large variation in the definition of urban followed by different countries. The low level of urban of urbanisation in India has been attributed to definitional specificity. One of the notable features of India's urban definition is that it has been stringent particularly the criterion of 75 per cent male main workers in non-agricultural sector compared to many countries including the countries of Southern Asia. If we drop this criterion then India's level of urbanisation rises to 44 per cent (Dijkstra et al., 2020). Alternatively based on an Agglomeration Index using population density (150 people per square kilometre), the minimum size of a large urban centre (50,000 inhabitants), and travel time to that urban centre (60 min) shows India to be 52 per cent urbanised (World Bank, 2013).

It is true that urban definition followed in India underestimates the degree of urbanisation. As such international comparison is not appropriate. However, retaining the same definition helps to know the tempo of urbanisation (i.e. per cent change in the degree of urbanisation). It also helps in understanding the regional disparities and rural and urban divide and aids in urban and development policy formulations.

India has experienced an accelerated economic growth after the central government launched economic reforms in 1991. The economic reforms aimed at loosening the control of the government and encouraged entrepreneurs to actively participate in India's economic development.

The economic growth reached about 8 per cent per annum during the first decade of the new millennium compared to a growth rate of about 3 per cent in the early 1980s. This has also led to a very spectacular change in the perception of the central government about urbanisation (Ahluwalia, 2011). In the Eleventh Five-Year Plan (2007–2012), it is argued that urbanisation should be seen as a positive factor in overall development. This change in the thinking is coincidental with the fact that urban areas presently contribute about three-fifth of the GDP, and a realisation that an ambitious goal of 9 to 10 per cent growth in GDP cannot be achieved without vibrant urban sector (Planning Commission, 2008). India also prepared the Twelfth Five-Year Plan (2012–2017), considering urban transition one of the major challenges which require a massive expansion in urban infrastructure and services. However, the Planning Commission was scraped by the central government in 2014 and replaced by NITI (National Institution for Transforming India) Aayog. In spite of change in the institutional structure at the apex level, the ideology of neo-liberalisation accompanied by privatisation and corporatisation of infrastructure and services in cities continued. The earlier programmes like JNNURM (Jawaharlal Nehru National Urban Renewal Mission) launched in 2005 have been changed to SMART Cities Mission and AMRUT (Atal Mission for Rejuvenation and Urban Transformation) in 2015. This chapter presents an assessment of the emerging pattern of urbanisation, its spatial pattern and the components of urban growth namely the contribution of natural increase, classification of rural into urban areas and the contribution of rural to urban migration. Although limited by the non-availability of 2021 census data, this chapter employs demographic techniques of analysis pertaining to the level, trends and decomposition of the urban growth using projected urban population and urbanisation by National Population Commission based on 2011 census data. This chapter elucidates the demographic dynamics of urbanisation in India relevant to its future development policies.

## 3.2 Trends in Urbanisation

According to the 2011 census, the urban population grew from 286 million in 2001 to 377 million in 2011—an increment of 91 million, and rural population increased from 742.5 million in 2001 to 833 million in 2011—an increment of 90.5 million. The rural population increment was lower than the urban increment for the first time since the 1951 census. The 2011 census also showed 19 million more than what was projected for the year 2011 showing an accelerated rate of urbanisation during 2001–2011. However, it must not be construed that projection methodology is not robust, but the transition of the urban trajectory is dependent upon a large of number factors associated with economic transition and rural to urban transformation.

Table 3.1 shows that urban population in India was 62 million in 1951 which constituted about 17 per cent of the total population. While urban population increased by 6 times during 1951 to 2011, the degree of urbanisation hardly doubled from 17 per cent in 1951 to 31 per cent in 2011. In terms of annual growth, the average was 2.32 per cent per annum during 1951–1961 which accelerated up to 3.79 per cent during 1971–1981, i.e. the highest urban growth since independence, after that it consistently decelerated until the 2001 census. During 2001–2011 the declining growth rate has seen a reversal as the urban population grew at the rate of 2.76 per cent per annum compared to 2.75 per cent per annum during 1991–2001. On the other hand, the degree of urbanisation in the country as a whole increased from 27.7 per cent in 2001 to 31.1 per cent in 2011—an increase of 3.3 percentage points during 2001–2011 compared to an increase of 2.38 percentage points in 1981–1991 and 2.1 percentage points during 1991–2001. Either the urban growth or absolute increase in the degree of urbanisation per se does not tell us about urbanisation rate. Urbanisation rate depends upon the tempo (also known as speed) of urbanisation as urbanisation it is influenced by urban–rural growth differentials. The tempo of urbanisation could be measured either through annual change in per cent urban population or through urban–rural growth differentials as shown in Table 3.2. The decelerated tempo of urbanisation has seen a significant turnaround during the decade 2001–2011 by both the indicators. When tempo of urbanisation increases it is natural that the urban growth rate would be higher than the rural growth rate and urban–rural growth differentials would be widening. Table 3.2 shows that the urban–rural growth differentials increased from about 1 per cent per annum during 1991–2001 to 1.61 per cent per annum during 2001–2011—a clear trend of the reversal in the deceleration of urbanisation observed in the last two decades.

However, the projected urban population shows that the turnaround urbanisation rate witnessed in the 2011 census may not continue in the following decades as tempo

**Table 3.1** Trends in urbanisation in India, 1951–2011

| Census year | Number of UAs/Towns* | Urban Population (in millions) | Per cent Urban | Average Annual Exponential Urban Growth Rate (%) |
|---|---|---|---|---|
| 1951 | 2843 | 62.44 | 17.29 | |
| 1961 | 2365 | 78.94 | 17.97 | 2.34 |
| 1971 | 2590 | 109.11 | 19.91 | 3.24 |
| 1981 | 3378 | 159.46 | 23.34 | 3.79 |
| 1991 | 3768 | 217.18 | 25.72 | 3.09 |
| 2001 | 4378 | 286.12 | 27.86 | 2.75 |
| 2011 | 6171 | 377.10 | 31.16 | 2.76 |

*Notes* As the 1981 census was not conducted in Assam, and the 1991 census was not held in Jammu and Kashmir, the population of India includes their projected figures
*Source* Figures up to 1991 are based on Census of India (1991); Data for 2001 and 2011 are based on Census of India 2001, A-04: Towns and urban agglomerations classified by population size class in 2001 since 1901 (PC01_A04_0); Census of India 2011, A-04: Towns and urban agglomerations classified by population size class with variation between 1901 and 2011 (PC11_A04-I); https://censusindia.gov.in/census.website/data/census-tables#, accessed on 13 November 2024
*The number of UA is based on all towns/outgrowths falling in a UA is counted as one. There were 474 UA and 5697 towns in 2011 census comprising a total of 6171 UAs/Towns

**Table 3.2** Urban–rural population growth rates and tempo of urbanisation, India, 1951–2011

| Decade | Rural (Average Annual growth rate* %) | Urban (Average Annual growth rate* %) | Urban–Rural Average Annual Growth Differentials (%) | Tempo of Urbanisation** (%) |
|---|---|---|---|---|
| 1951–1961 | 1.87 | 2.34 | 0.47 | 0.37 |
| 1961–1971 | 1.97 | 3.24 | 1.27 | 1.07 |
| 1971–1981 | 1.76 | 3.79 | 2.03 | 1.72 |
| 1981–1991 | 1.80 | 3.09 | 1.29 | 1.01 |
| 1991–2001 | 1.69 | 2.75 | 1.06 | 0.83 |
| 2001–2011 | 1.15 | 2.76 | 1.61 | 1.20 |

*Source*: Same as in Table 3.1
* Exponential growth rate
**Average annual change in per cent urban

of urbanisaton is likely to slow down as per the given definition of urban followed by the Census of India. Table 3.6 also shows a rapid decline in rural population growth reaching 0.27 per cent per annum by 2031. It is important to recognise that slowing down of urbanisation rate indicates that India does not follow the historical model of urbanisation based on Western experience as enunciated by two-sector model of Lewis (1954). It is assumed in the model that there is an unlimited supply of labour in the traditional agrarian sector which will be transferred to modern capitalist sector as economic growth takes place. This process continues until unlimited labour supply ceases to exist resulting in rising capital accumulation and ploughing back spurring increasing labour productivity, rural to urban migration and urbanisation. The Lewisian expectation of rural–urban transformation by many scholars is belied in Indian situation due to following reasons:

i) Dependence on agriculture is still very high. About half of the workforce is dependent on agriculture contributing to just 14 per cent of India's GDP (McKinsey and Company and Confederation of Indian Industry, 2013).
ii) A significant proportion of rural to urban migration is not of permanent nature. There is a significant short-term seasonal, temporary and circulatory migration taking place as a livelihood strategy (Bhagat & Keshri, 2023).
iii) Urbanisation manifests not in the form of rural–urban transformation but in the form of rural to urban transition. Urban transition in India and many developing countries is a complex process influenced by the interwoven nature of economic transition and demographic change which is neither dichotomous in the sense of exclusive urban or rural nor linear in terms of sectoral transformation.

In India, the rural-urban classification should attempt to identify urban, peri-urban and semi-urban area (transitional areas). However, a complete overhaul in the urban definition is not required as that would make the historical data on India's urbanisation redundant. A tweaking of the share of non-agricultural workforce may be suffice to identify semi-urban areas. For example, non-agricultural workforce above 50 per cent but lower than 75 per cent may be taken as an indicator of semi-urban areas along with other demographic indicators of size and density. On the other hand, adjunct towns as a part of the UA including census towns and outgrowths combined with semi-urban areas may be identified as peri-urban areas.

## 3.3 State Level Patterns

At the state level, the pattern of urbanisation is very diverse, but economically advanced states show a high level of urbanisation. The earlier studies show that the per capita income and per cent urban is strongly positively correlated at the state level (Bhagat & Mohanty, 2009) which is also found true for census year 2011. All southern states and northern states like Punjab, Haryana and western states namely Gujarat, Maharashtra and the eastern state of West Bengal have higher urbanisation level than the national average. The small state of Goa continues to top the list of states with 62 per cent urban population and also has highest per capita income among the states (except Delhi which is like a city state) (Ministry of Finance, 2012). Among major states, Tamil Nadu continues to be ahead of other states with 48.4 per cent level of urbanisation in 2011. The states which are at the bottom of urbanisation are also the states having lower level of income than the national average. These states are Bihar (11.3), Assam (14 per cent), Orissa (16.6) UP (22 per cent), Jharkhand (24 per cent) and Rajasthan (24 per cent). Himachal Pradesh—a hilly state is an exception with higher level of income but with lowest level of urbanisation among states in India (Fig. 3.1).

Although reversal in the deceleration of urbanisation at the national level is a major feature, there are only 15 states and UTs out of 35 that showed increased urban population growth rate during 2001–2011 compared to 1991–2001. Among them, Kerala, Andhra Pradesh, Karnataka, Gujarat, West Bengal, Bihar, Jharkhand,

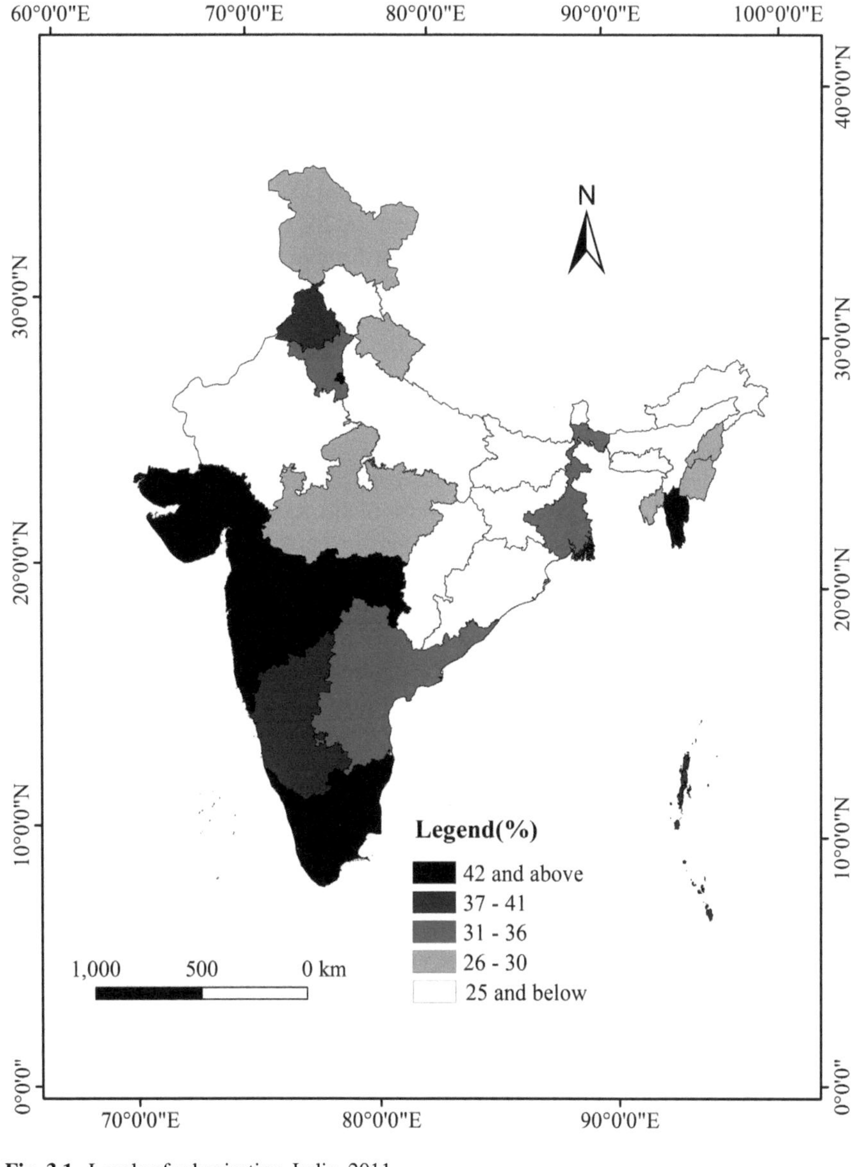


**Fig. 3.1** Levels of urbanisation, India, 2011

Chhattisgarh and Uttarakhand are the major states. A very high urban population growth has occurred in the states of Kerala and Andhra Pradesh where urban population growth rate has increased to 6.5 per cent per annum in Kerala and 3 per cent per annum in Andhra Pradesh during 2001–2011 compared to just about 1 per cent per annum during 1991–2001. In both Kerala and Andhra Pradesh along with West Bengal and Gujarat, a large number of census towns have emerged as a result of rural–urban classification in 2011 (Bhagat 2011).

It is worthwhile to mention that urbanisation has occurred very fast in the state of Kerala where urban population growth increased from less than 1 per cent per annum during 1991–2001 to 6.5 per cent per annum during 2001–2011. The state has seen the emergence of 461 new census towns emerged as a result of rural–urban classification in 2011 compared to 99 new census towns in 2001. Most of the new census towns have emerged as a part of the urban agglomerations of the existing big cities leading to the formation of six new million plus UAs in 2011 in Kerala. In 2001, there was only one million plus UA, i.e. Kochi; in 2011, Thiruvananthapuram UA, Kozhikode UA, Thrissur UA, Malappuram UA, Kannur UA, Kollam UA have been added to the list of million plus UAs in Kerala. As a result, the level of urbanisation of the state increased from 26 per cent in 2001 to 47 per cent in 2011. This also shows the emerging spatial dimension of urbanisation which is consistent with massive infrastructure development of transport, communication, real estate and dispersal of industries in the last decade which brought about changes in many villages to qualify into urban areas.

Various studies show that urbanisation has been closely related to economic development, and is the single most important factor in the organisation of production and access to services. Figure 3.2 shows the relationship between per capita net domestic product (i.e. per capita income) at the state level and degree of urbanisation. There is a high positive significant correlation between the two variables (+0.88). It also explains about 77 per cent variations in the degree of urbanisation across the states of India. On the other hand, Fig. 3.3 shows the relationship between per capita income and urban–rural growth differentials which is the basis of the tempo of urbanisation. The correlation co-efficient is positive, i.e. +0.68 and explains 46 per cent variations in the urban–rural growth differentials. This shows that the rising income is positively related with both the degree and tempo of urbanisation, although the impact on tempo is lower than the degree of urbanisation. As the explained variation in the urban–rural growth differentials is low, this shows that the faster economic growth during the 2000s (about 8 per cent growth of GDP at the national) alone cannot explain the differential pattern of urbanisation, and much more depends upon how rural areas have been transformed into urban and how urban agglomerations are emerging along with outgrowths (OGs) adjacent to the cities and towns which are spatially connected and functionally dependent. The convergence of economic and spatial transitions is very important for the dispersal of economic growth and sustainable development.

As stated earlier, urbanisation is the product of urban–rural growth differentials. Table 3.3 presents urban–rural growth differentials along with urban growth rate and degree of urbanisation (% urban). To conclude this section, we may reiterate that there exists a positive relationship between per capita income and degree of

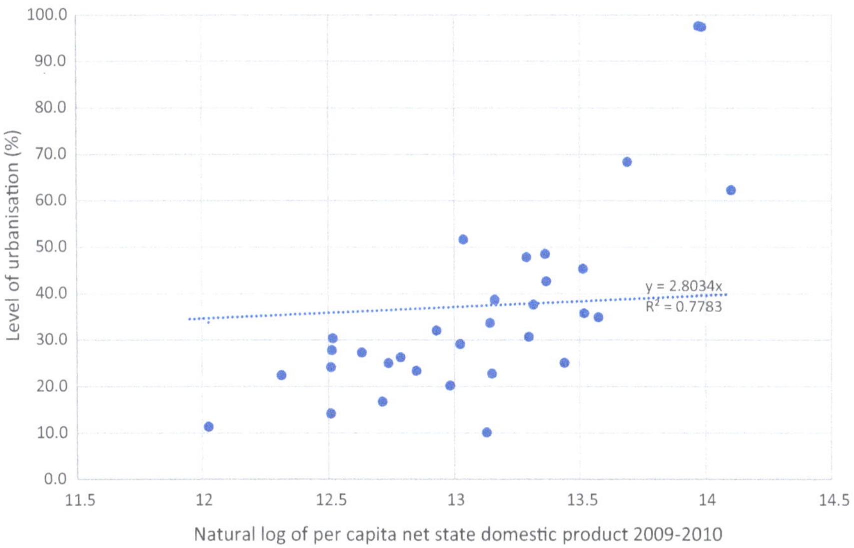


**Fig. 3.2** Relationship between per capita income of State/UT and level of urbanisation, 2011 (N = 32)

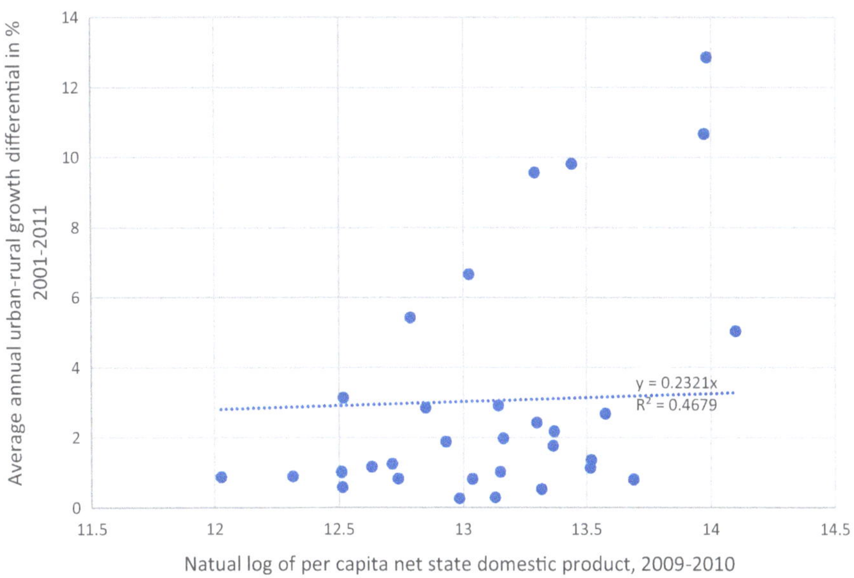


**Fig. 3.3** Relationship between per capita income of State/UT and urban–rural growth differentials, 2001–2011 (N = 32)

urbanisation and it further shows that the level of economic development contributes positively in widening urban–rural growth differentials and thus contributing to the tempo of urbanisation. In the next section, we would like to assess the demographic sources of urban population growth to understand the role of economic development in the process of urbanisation.

## 3.4 Demographic Sources of Urban Population Growth

Urban population growth is the product of several factors namely (i) natural increase, (ii) net rural to urban migration, (iii) net rural–urban reclassification and (iv) Jurisdictional changes/or changes in the municipal boundaries. Their respective contributions of the demographic sources to urban growth depend upon rate of economic growth influencing demand of labour and rural to urban migration, demographic changes like fertility and mortality decline, urban and regional development policies, growth and expansion of infrastructure, national definitions of urban space, etc.

**Natural increase:** Urban natural increase is an important factor in the growth of urban population results from an excess of births over deaths in urban areas. It depends on levels of birth rate (the number of births per 1000 population) and death rate (the number of deaths per 1000 population). Both birth and death rates depend upon age sex structure of the population. The birth and death rates are generally lower due to early decline of birth and death rates in urban compared to rural areas. However, despite a lower level of birth rate, there is often an excess of births over deaths in urban areas due to a lower level of mortality and a younger age distribution resulting from migration of youth to urban areas.

**Net rural to urban migration:** Internal migration to urban areas from rural areas or international migration contributes to urban growth. In the case of India, international migration plays an insignificant role in urban growth. However, immigrants may be significant at some city level, but the size of emigrants is not known. This is a significant limitation in the estimation of net rural to urban migration. Generally urban areas in India are growing by internal migration of rural to urban stream. A net rural to urban migration means rural to urban migration minus urban to rural migration. There is a very small component of rural to urban unclassifiable and this is generally excluded from the rural–urban streams. As such this underestimates the contribution of net rural to urban migration marginally (Bhagat & Mohanty, 2009).

**Net Rural–urban reclassification:** This component refers to the extent to which the rural areas designated earlier so get classified into urban areas based on the application of accepted urban definition. Further, the municipal or boundaries of the existing cities and towns may also be redrawn bringing the rural areas into the urban fold. The entire process is characterised and known as in situ urbanisation. This process occurs due to demographic and economic transition of existing rural settlements rather than rural to urban migration. In India, rural and urban classification is a dynamic process which means that rural areas are classified into urban as well as urban areas classified may also be declassified and put under the category of rural.

**Table 3.3** Degree of urbanisation and urban growth in India and States, 2001–2011

| State/India | % Urban 2001 | % Urban 2011 | Average Annual Urban Growth Rate* | Average Annual Rural Growth Rate* | Urban–rural Annual Growth Differentials* | Tempo of Urbanisation (Annual Change in Per cent Urban)* |
|---|---|---|---|---|---|---|
| Andhra Pradesh | 27.4 | 33.4 | 3.09 | 0.19 | 2.90 | 2.18 |
| Arunachal Pradesh | 20.8 | 22.6 | 3.18 | 2.07 | 1.01 | 0.86 |
| Assam | 12.9 | 14.0 | 2.43 | 1.41 | 1.02 | 0.85 |
| Bihar | 10.5 | 11.3 | 3.01 | 2.15 | 0.86 | 0.76 |
| Chhattisgarh | 20.1 | 23.2 | 3.49 | 1.65 | 2.84 | 1.54 |
| Goa | 49.8 | 62.1 | 3.01 | –2.02 | 5.03 | 2.46 |
| Gujarat | 37.4 | 42.5 | 3.06 | 0.89 | 2.17 | 1.36 |
| Haryana | 29.0 | 34.7 | 3.66 | 0.99 | 2.67 | 1.96 |
| Himachal Pradesh PradeshPradesh | 9.8 | 10.0 | 1.45 | 1.17 | 0.28 | 0.2 |
| Jammu & Kashmir | 24.9 | 27.2 | 3.04 | 1.88 | 1.16 | 0.92 |
| Jharkhand | 22.2 | 24.0 | 2.79 | 1.79 | 1.00 | 0.81 |
| Karnataka | 34.0 | 38.5 | 2.72 | 0.75 | 1.97 | 1.32 |
| Kerala | 25.9 | 47.7 | 6.56 | –3.00 | 9.56 | 8.41 |
| Madhya Pradesh | 26.7 | 27.6 | 2.28 | 1.70 | 0.58 | 0.33 |
| Maharashtra | 42.4 | 45.2 | 2.12 | 0.99 | 1.13 | 0.66 |
| Manipur | 24.1 | 30.2 | 3.55 | 0.43 | 3.12 | 2.53 |
| Meghalaya | 19.6 | 20.0 | 2.7 | 2.45 | 0.25 | 0.2 |
| Mizoram | 49.4 | 51.5 | 2.42 | 1.61 | 0.81 | 0.42 |
| Nagaland | 17.2 | 28.9 | 5.15 | –1.5 | 6.65 | 6.8 |
| Orissa | 15.0 | 16.6 | 2.37 | 1.13 | 1.24 | 1.06 |
| Punjab | 34.0 | 37.4 | 2.28 | 0.76 | 0.52 | 1 |
| Rajasthan | 23.4 | 24.8 | 2.56 | 1.74 | 0.82 | 0.59 |
| Sikkim | 11.0 | 24.9 | 9.29 | –0.52 | 9.81 | 1.26 |
| Tamil Nadu | 44.2 | 48.4 | 2.4 | 0.64 | 1.76 | 0.95 |
| Tripura | 17.1 | 26.1 | 5.65 | 0.23 | 5.42 | 5.26 |
| Uttar Pradesh | 20.8 | 22.2 | 2.52 | 1.64 | 0.88 | 0.67 |
| Uttarakhand | 25.6 | 30.5 | 3.49 | 1.07 | 2.42 | 1.91 |
| West Bengal | 27.9 | 31.8 | 2.61 | 0.74 | 1.87 | 1.39 |
| Andaman & Nicobar Islands | 32.6 | 35.6 | 1.53 | 0.18 | 1.35 | 0.91 |
| Chandigarh | 89.7 | 97.2 | 1.3 | –11.55 | 12.85 | 0.83 |

(continued)

**Table 3.3** (continued)

| State/India | % Urban 2001 | % Urban 2011 | Average Annual Urban Growth Rate* | Average Annual Rural Growth Rate* | Urban–rural Annual Growth Differentials* | Tempo of Urbanisation (Annual Change in Per cent Urban)* |
|---|---|---|---|---|---|---|
| Dadra & Nagar Haveli * | 22.8 | 46.6 | 11.52 | 0.73 | 10.79 | 10.4 |
| Daman & Diu | 36.2 | 75.1 | 11.58 | –5.12 | 16.70 | 10.7 |
| Delhi | 93.6 | 97.5 | 2.35 | –8.31 | 10.66 | 0.41 |
| Lakshadweep | 44.5 | 78.0 | 6.23 | –8.68 | 14.91 | 7.52 |
| Pondicherry | 66.6 | 68.3 | 2.71 | 1.91 | 0.80 | 0.25 |
| **India** | **27.8** | **31.1** | **2.76** | **1.15** | **1.61** | 1.2 |

*average annual during 2001–2011
*Source* Census of India 2011, PCA-SD: Primary Census Abstract (PCA) Data, India and States/UTS-State and District Level, 2011, https://censusindia.gov.in/nada/index.php/catalog/6191; Census of India 2001, PCA-TOT: Primary Census Abstract, Total, https://censusindia.gov.in/census.website/data/census-tables, accessed on 15 November 2024

That's why the net rural to urban classification is usually used in the Indian context to estimate the contribution of this factor.

Several studies have shown that natural increase plays a very dominant role in India's urbanisation (Bhagat & Mohanty, 2009; Visaria, 1997). The trend in the natural increase for four decades up to 2010 is presented in Table 3.4. The natural increase in urban areas remained at 19.3 per 1000 persons during 1970–1980 which declined to 13.2 during 2001–2010. On the other hand, natural increase in rural areas declined from 20 per 1000 population during 1971–1980 to 17.3 during 2001–2010—a decline of just 3 points compared to decline of 6 points in urban areas. Due to faster decline of natural increase in urban areas the urban–rural growth differentials have also widened during the last four decades. This was evident in the fact that almost no urban–rural differential in natural increase was seen during the 1970s; however, it increased to 2 per 1000 population during the 1980s and further increased to about 4 per 1000 during the 1990s, remained constant thereafter in the 2000s, and has shown a slight decline in the 2010s (see Table 3.4). The differential natural increase between urban and rural areas is not only the result of fertility differentials but mortality differentials as well. Mortality decline has been faster in rural areas leading to a widening urban–rural differentials in natural increase until 2002 showing a potential for rural and urban migration, but to what extent this potential is realised depends upon the economic and occupational transition. In India, fertility started declining since the early 1970s. The onset of fertility decline was early and faster in urban than rural areas. In a situation of widening urban–rural growth differentials in natural increase, the other components like net rural–urban classification of settlements and net rural to urban migration need to have a larger share in order to first compensate the deficit in urban population growth arising due to faster decline in natural increase in urban areas. The lower natural increase in urban compared to rural areas is compensated by rural to urban migration and as well as rural to urban settlement classification, however, in order to accelerate urbanisation the latter two components must increase at a higher rate to push up the proportion of population living in urban areas. In a situation of declining urban natural increase, the contribution of net rural to urban classification (new towns minus old towns reverted to the status of rural) and net rural to urban migration (rural to urban stream of migration minus urban to rural stream of migration) would be decisive in the process of urbanisation (Bhagat & Mohanty, 2009).

As there has been an increasing differential in natural increase between urban–rural areas, until 1990s and remained stable during 2000s, it may be concluded that the declining natural increase in urban areas to some extent negatively affected the tempo of urbanisation during 2001–2011, while the turnaround in urbanisation during the 2000s was mainly contributed by the classification of rural into urban areas and rural to urban migration. Among the two factors, namely rural–urban classification and rural to urban migration, the rural to urban classification which also includes changes in the municipal boundaries of the existing cities and towns and the reorganisation of urban areas into UA consisting of outgrowths (OGs) seems to be more dominant factor compared to rural to urban migration.

**Table 3.4** Birth, death and natural increase per 1000 population by rural and urban areas, 1971–2010, India

| Years | Average Birth rate (per 1000) | Average Death rate (per 1000) | Average natural increase (per 1000) | Urban–rural differentials in natural increase |
|---|---|---|---|---|
| 1971–1980 | | | | |
| Rural | 35.8 | 15.8 | 20.0 | |
| Urban | 28.5 | 9.2 | 19.3 | –0.7 |
| 1981–1990 | | | | |
| Rural | 33.9 | 12.6 | 21.3 | |
| Urban | 27.0 | 7.7 | 19.3 | –2.0 |
| 1991–2000 | | | | |
| Rural | 29.4 | 9.9 | 19.5 | |
| Urban | 22.3 | 6.5 | 15.8 | –3.7 |
| 2001–2010 | | | | |
| Rural | 25.7 | 8.4 | 17.3 | |
| Urban | 19.3 | 6.0 | 13.2 | –4.1 |
| 2011–2020 | | | | |
| Rural | 22.2 | 7.0 | 15.2 | |
| Urban | 17.0 | 5.4 | 11.6 | –3.4 |

*Source* Sample Registration System, Various Years, Registrar General and Census Commissioner, India (www.censusindia.gov.in)

The precise contribution of these components of urban growth is presented in Table 3.5. The contribution of natural increase in the total urban population increment was 44 per cent during 2001–2011 compared to 58 per cent in the previous decade. On the other hand, the contribution of net reclassification of rural to urban areas including changes in municipal boundaries and outgrowths (OGs) have increased very significantly from about 22 per cent during 1991–2001 to about 36 per cent during 2001–2011. This factor has been dominant in influencing the speed of urbanisation during the 2000s compared to net rural to urban migration. Although net rural to urban migration has increased 14.2 million to 20.7 million, the net rural to urban classification added a population of 30.4 million during 2001–2011 compared to 14.7 million during 1991–2001. The 2011 census reported that the number of towns at the national level increased from 5161 to 7935—a net addition of 2774 towns[1] (2532 census towns and 242 statutory towns) in 2011 compared to the net additions of 763 and 693 towns in the 1991 and 2001 census, respectively. A four-fold increase of new towns mostly small towns (less than 20,000) shows the overriding importance of spatial changes that reorganised the rural–urban space and produced faster urbanisation during the 2000s. A significant number of these new small towns have emerged in more developed states. For example, about 44 per cent of new towns have emerged in only four states namely West Bengal, Kerala, Tamil Nadu and Maharashtra comprising of little more than 50 per cent of the total population added by all new towns in India in the 2011 census. This shows how regional disparities are being shaped by in situ *urbanisation* and why there is a need of an urbanisation policy to disperse the impulse of urbanisation in more underdeveloped areas of the country.

It is very important to highlight that the three components of urban growth namely, natural increase, net reclassification of settlements and net rural to urban migration are not independent but mutually reinforcing each other. The history of demographic transition in Western countries informs us that migration followed the increasing natural increase due to mortality decline and induced fertility decline consequently. On the other hand, in India the spurt in the contribution of reclassification is not independent of migration. The nature of urbanisation purported by reclassification is known as in situ urbanisation. A large number of people also migrate from smaller villages to large and economically diversifying villages enabling them to be qualified as urban when census criteria are applied in the rural–urban classification. It is also important to realise that the components of the sources of urban population growth namely natural increase, rural to urban classification, rural to urban migration and jurisdictional changes are not neatly independent but mutually interdependent driving urban transition and urbanisation rate.

### *3.4.1 Projected Trend in Urbanisation*

The projection of urban population is a two-stage process. First the total population of a country or region/province by age groups and sex is projected. In case of India and states Component Method of projections were used assuming that the inter-state

**Table 3.5** Contribution of the components of urban growth, India, 1971–2011

| Components | Population in Million | | | | Percentage Distribution | | | |
|---|---|---|---|---|---|---|---|---|
| | 1971–1981 | 1981–1991 | 1991–2001 | 2001–2011 | 1971–1981 | 1981–1991 | 1991–2001 | 2001–2011 |
| Urban increment | 49.9 | 56.8 | 68.2 | 91.0 | 100.0 | 100.0 | 100.0 | 100.0 |
| Natural increase (of initial population plus inter-censal migrants) | 24.9 | 35.4 | 39.3 | 39.9 | 50.0 | 62.3 | 57.6 | 43.8 |
| Net rural–urban migration | 9.3 | 10.6 | 14.2 | 20.7# | 18.6 | 18.7 | 20.8 | 22.7 |
| Net reclassification from rural to urban including jurisdictional changes | 15.7 | 10.8 | 14.7 | 30.4 | 31.4 | 19.0 | 21.5 | 33.4 |

*Source* Figures up to 1991–2001 are taken from Bhagat and Mohanty (2009). The components of 2001–2011 are estimated based on natural increase in urban areas between 2001–2010 using exponential growth rate

*Note:* The final figure of net reclassification (i.e. new towns—declassified town) is based on information in Table A04 Appendix for the 2011 census. The impact of boundary changes during 2001–2011 appears negligible as it is driven by urbanised area. It may be noted that the estimates of net rural to urban classification for the period 2001–2011 published in earlier studies were based on provisional data from census 2011, and net rural to urban migration was estimated residually (Bhagat, 2012, 2016)

#This is based on final migration data based on D-2 Migration Tables, Census of India 2011. The net rural to urban migration (i.e. Rural to Urban minus Urban to Rural) with duration of 0–9 years was 18.4 million. However, it was adjusted for rural–urban unclassifiable (11.7 per cent of total internal migrants), and duration not stated (6.1 per cent for rural to urban and 13 per cent for urban to rural migrants). The adjusted figure for net rural to urban migration was 20.7 million during 2001–2011

**Table 3.6** Projected urbanisation and urban growth in India, 2021 to 2031

| Year | Urban Population (in millions) | Per cent Urban | Annual Exponential Rural Growth Rate (%) | Annual Exponential Urban Growth Rate (%) | Urban–rural growth differentials | Tempo of urbanisation (annual change in per cent urban) |
|---|---|---|---|---|---|---|
| 2011 | 377.10 | 31.16 | 1.15 | 2.76 | 1.61 | 1.2 |
| 2021* | 469.92 | 34.52 | 0.66 | 2.20 | 1.54 | 1.07 |
| 2031* | 558.87 | 37.88 | 0.27 | 1.73 | 1.46 | 0.97 |

*Projected figures, see National Population Commission (2019)

net migrants during decade 2001–2011 remained constant throughout the projection periods. It is further assumed that the international migration remained to be negligible. Thus, it has not been considered in the projection exercise. In the second stage, proportion of urban population was projected based on the URGD (urban–rural growth differentials) method as given in Manual—VIII of U.N. (United Nations, 1974). This method is based on the fact that the urban–rural growth differentials follow a logistic pattern. The URGD for the period 2001–2011 has been assumed to be same in the projection period. The projected urban population and rural and urban growth rates for India are provided for the year 2021 and 2031 in Table 3.6.

The projected urban population shows a decline in urban growth rate to 2.20 per cent per annum during 2011–2021 compared to 2.76 per cent during 2001–2011 and further declines to 1.73 per cent during 2021–2031. The projected urban population also shows that the degree of urbanisation has reached 34.5 per cent in 2021, and shall be 37.8 per cent in 2031. It shows an increase of 3.4 per centage points during 2011–2021 and 3.3 percentage points during 2021–2031—a very stable increase of 3 percentage points since 2011, however, the tempo of urbanisation has been declining either measured through urban–rural growth differentials or annual change in the per cent of urban population (see Table 3.6). However, as shown in Fig. 3.4 based on decomposition exercise on the sources of urban population growth for the decade 2011–2021, a substantial part of the components of urban growth continues to result from rural–urban classification, i.e. emergence of census towns, extension of the municipal boundaries and delimitation of UA owing to the urban sprawl similar to previous decade. This vindicates the emerging fact of earlier studies that in situ *urbanisation* has been a significant mode of rural to urban transition in India.

## 3.5 Concluding Remark

A stringent definition of urban areas keeps India's urbanisation at a low level. However, the changing definition of urban will make more than a centuries' data on urbanisation redundant. The utility of data on urbanisation lies not in the trend per se but what are the sources of this change and how the relational urban space

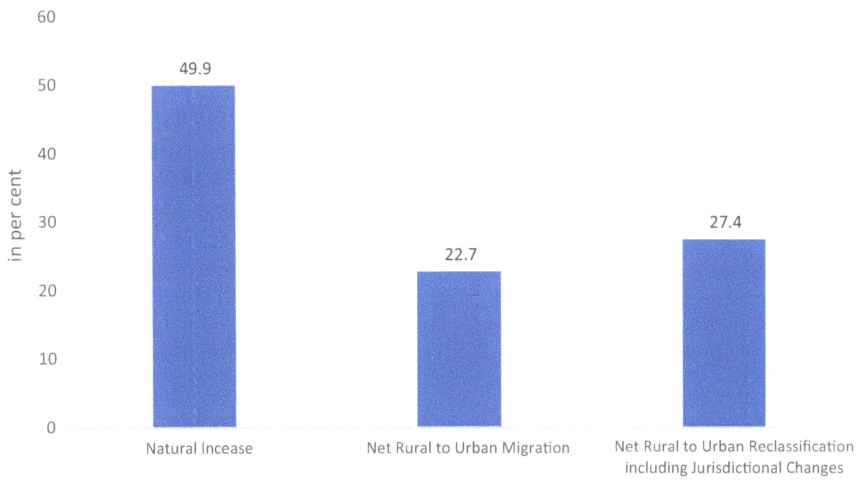


**Fig. 3.4** Components of projected urban population growth during 2011–2021, India (in %) (*Source* by Authors; Natural increase is estimated based on Rate of Natural Increase given in Table 3.4 using exponential growth rate. Net rural to urban migration is assumed to be constant and Net Rural to Urban Classification including jurisdiction changes is estimated residually)

has been changing to understand the trend in inter-regional disparities or inter-state disparities and rural and urban divide in India. To make this argument clear, we may state the fact that as per present definition, Bihar's degree of urbanisation is one-third of India. However, changing the urban definition is not likely to change the relational urban space. Further, 45 per cent of India's workforce is dependent on agriculture contributing to 14 per cent of GDP. The rural–urban transformation through definitional changes alone is not likely to bring sectoral transformation of India's economy.

The rate of urbanisation reached a peak during the 1970s, and witnessed a declining tempo during the 1980s and 1990s. It is although reversed during the 2000s, but the projected urbanisation rate shows a slowing down. An important feature of India's urbanisation emerging from the analysis of census data is that the urban transition is the result of rural–urban classification rather than increasing trend in rural to urban migration. This may be better called in situ urbanisation characterised by the increasing reclassification of rural into urban areas, changes in the jurisdiction of municipal boundaries and the delimitation of urban agglomerations and designation of settlements or part of a settlement as outgrowths (OGs) generally adjacent to the cities and large towns which are spatially connected and functionally dependent.

Thus, the geographical expansion of urbanisation through the emergence of census towns and spill over from the existing big cities to its peripheral areas need to be recognised in our urbanisation policy. It will be good to strengthen the form of urbanisation which is spatially distributed and led by a large number of medium and small urban centres because of their potential to influence rural economic growth and interdependent rural and urban relationship. Unfortunately, India's urban policy is dominated by urban development policies focused on large cities.

This raises an important issue of providing infrastructure, civic amenities and improving governance in large number of small and medium urban centres. There is a need to make a distinction between an urban development policy and an urbanisation policy focusing on medium and small urban centres aiming to achieve rural to urban transformation and envisioning urbanisation as a vehicle of rural development.

**Acknowledgements** This is to acknowledge that the chapter is a revised version of the working paper by the first author entitled 'Urbanisation in India: Trend, Pattern and Policy Issues', Working Paper No. 17, 2018, International Institute for Population Sciences, Mumbai, https://iipsindia.ac.in/sites/default/files/IIPS_Working_Paper_No_17.pdf, accessed on 10 November 2024.

## Notes

1. A large number of census towns emerged in the 2011 census have been attributed to census activism (Kundu 2011). In actual practice, however, the list of new census towns is prepared by the Directorate of Census Operation at the state level as per the circular by the Census Commissioner and Registrar General, India. Directorates have been advised to identify urban places as per definition of urban adopted in the 2011 census considering all rural places having a population of 4000 or more at the 2001 census. It is assumed that rural places having population of 4000 in the 2001 census is expected to cross 5000 population mark during the 10 years by the time of the 2011 census. Further, the circular issued by the Census Commissioner and Registrar General does not show any activism by the census in identification of census towns rather it shows the utmost sincerity of the census 2011 in classification of places by rural and urban status in view of its growing significance in planning, development and politics of the country (See Census of India 2011, No.2/1/2008-SS, Circular No. 2, dated 23.07.2008).

## References

Ahluwalia, M. (2011). Prospects and Policy Challenges in the Twelfth Plan. *Economic and Political Weekly, 46*(21), 88–105.

Bhagat, R. B. (2005). Rural-Urban Classification and Municipal Governance in India. *Singapore Journal of Tropical Geography, 26*(1), 61–74.

Bhagat, R. B. (2011). Emerging Pattern of Urbanisation in India. *Economic and Political Weekly, 46*(34), 10–12.

Bhagat, R. B. (2012). A Turnaround in India's Urbanization. *Asia Pacific Population Journal, 27*(20), 23–39.

Bhagat, R. B. (2016). Changing Pattern of Migration in India. In C. Guilmoto & G. Jones (Eds.), *Contemporary Demographic Transformations in China, India and Indonesia* (pp. 239–254). Springer.

Bhagat, R., & Keshri, K. (2023). Internal migration and temporary labour circulation in India. In K. James & T. Sekher (Eds.), *India population report* (pp. 246–269). Cambridge University Press. https://doi.org/10.1017/9781009318846.009

Bhagat, R. B., & Mohanty, S. (2009). Emerging Pattern of Urbanisation and Contribution of Migration in Urban Growth in India. *Asian Population Studies, 5*(1), 5–20.

Census of India. (1991). Emerging trends of urbanisation in India, Occasional Paper No. 1 of 1993, prepared by Jain, M.K., Ghosh, M. and Kim, W.B., Office of the Registrar General, and Census Commissioner.

Champion, T. (1999). Urbanisation and Counterurbanisation. In M. Pacione (Ed.), *Applied Geography: Principles and Practice* (1st ed., pp. 347–357). Routledge. https://doi.org/10.4324/9780203012512

Codrington, S. (2005). *Planet Geography.* Solid Star Press Sydney.

Cohen, B. (2004). Urban Growth in Developing Countries: A Review of Current Trends and a Caution Regarding Existing Forecasts. *World Development, 32*(1), 23–51.

Davis, K. (1962). Urbanisation in India: Past and future. In R. Turner (Ed.), *India's Urban Future* (pp. 3–26). University of California Press.

Dijkstra, L., Hamilton, E., & Lall, S. (2020). How Do We Define Cities, Towns, and Rural Areas? World Bank Blog. https://blogs.worldbank.org/sustainablecities/how-do-we-define-cities-towns-and-rural-areas, 10 March 2020, Accessed on 16 August, 2020.

Ellis, P., & Roberts, M. (2016). *Leveraging Urbanization in South Asia: Managing Spatial Transformation for Prosperity and Livability.* World Bank.

Kundu, A. (2011). Method in Madness: Urban data from 2011 Census. *Economic and Political Weekly, 46*(40), 13–16.

Lewis, A. (1954). Economic Development with Unlimited Supplies of Labour. *The Manchester School, 22*(2), 139–191.

McKinsey and Company and Confederation of Indian Industry. (2013). *India as an Agriculture and High Valued Food Powerhouse: A New Vision for 2013.* McKinsey and Company Inc.

Ministry of Finance. (2012). *Economic survey, 2011–12.* Govt. of India, New Delhi.

Planning Commission, Govt. of India (2008). *Eleventh Five Year Plan, Vol III: Agriculture, Rural Development, Industry, Services and Physical Infrastructure.* Oxford University Press.

National Population Commission. (2019). *Population Projections for India and States, Report of the Technical Group on Population Projections.* Census of India 2011 and Ministry of Health and Family Welfare.

United Nations. (1974). *Manual VIII: Methods for Projections of Rural and Urban Population* (Population Studies No. 55). Department of Economic and Social Affairs, United Nations.

United Nations. (2009). *The World Urbanization Prospects: The 2005 Revision.* Population Division, Department of Economic and Social Affairs.

United Nations, Department of Economic and Social Affairs, Population Division. (2019a). *World Urbanization Prospects: The 2018 Revision* (ST/ESA/SER.A/420). United Nations.

United Nations, Department of Economic and Social Affairs, Population Division. (2019b). *World Urbanization Prospects 2018: Highlights* (ST/ESA/SER.A/421).

Visaria, P. (1997). Urbanisation in India: An Overview. In *Urbanization in Large Developing Countries: China, Indonesia, Brazil, and India* (eds. Gavin Jones and Pravin Visaria). Clarendon Press.

World Bank. (2009). *World Development Report 2009: Reshaping Economic Geography.* The World Bank. https://doi.org/10.1596/978-0-8213-7607-2

World Bank. (2013). *Urbanization Beyond Municipal Boundaries: Nurturing Metropolitan Economies and Connecting to Peri-Urban Areas in India.* World Bank.

# Chapter 4
# Dynamics of Population Change by Size Class of Urban Centres

## 4.1 Introduction

The size of population of an urban centre is a significant attribute reflecting its economic and social conditions. The size may vary from a tiny town to a colossal city of several millions forming an urban continuum (Pacione, 2001). The increasing size helps an urban centre to grow benefitting from the increasing agglomeration economies linked to size, however there are costs to big cities as well due to rising congestion, pollution and diseconomies associated with overcrowding (Glaeser, 2010). As size increases the internal morphology of big cities also changes from a monocentric to a polycentric urban structure compared to small and medium size urban centres usually having a single business centre known as CBD (central business district). This is mostly due to the fact that the large urban centres grow faster compared to the small and medium urban centres in an evolving urban system, albeit this happens generally in the early phases of urbanisation. On the other hand, as congestion and crowding sets in the large urban centres, the smaller urban centres are likely to grow faster particularly in the latter phases of the urbanisation cycle. Geyer and Kontuly (1993) identified three phases in the urbanisation cycle namely the primate city,[1] intermediate city and small city phases. In early phases of urbanisation primate cities grow rapidly followed by polarisation reversal led by the higher growth rates of intermediate cities and declining growth rates of primate cities, and lastly in the phase of counter-urbanisation small cities show higher growth rates but declining growth rates of primate and intermediate cities. The prime reason for the differential population growth rates of urban centres is the net migration, and the entire process is known as differential urbanisation. Differential urbanisation is not a finite process but a cycle could repeat in varying degrees. On the other hand, counter-urbanisation may turn into reurbanisation again causing higher population growth rates in primate cities. It is thus to be noted that differential urbanisation is a cyclical process of concentration and deconcentration of population in human settlements defined as urban (Champion, 2001).

 
R. B. Bhagat and M. I. Hassan, *Urbanisation and Urban Policies in India*, Exploring Urban Change in South Asia, https://doi.org/10.1007/978-981-96-3181-0_4

There are no standard criteria to categorise cities and towns. It is important to underscore that whichever way urban settlements are categorised, size remains the important criterion of categorising cities and towns. The study of urban population growth by size class of urban centres could throw light on the nature of urban transition in a country and the growth differentials would show not only the magnitude of rural to urban migration but also urban to urban migration of inter-city/town nature. On the other hand, while urban natural increase continues to contribute significantly to the urban growth of India, its role in explaining differential growth rates of cities and towns may not be as large as the net rural to urban migration and net inter-city/town migration could play along with changes in the municipal boundaries (Bhagat, 2004). As data on city and town levels are limited, this chapter attempts to show only the differential population growth of urban centres by size class of cities and towns.

India launched a new economic policy in 1991 aiming to achieve higher economic growth expected to produce increased urban population growth rate in subsequent decades. The actual urban population growth rate, however, has slowed down in the decade 1991–2001 at the all-India level but has seen a turnaround during 2001–2011. It, therefore, is curious to know how large urban centres are growing compared to the small and medium urban centres. An attempt has been made in this chapter to analyse the urban population growth by size class of urban centres for India and different states.

## 4.2 Urban Population Growth by Size of Urban Centres: Some Measurement Issues

As also mentioned in Chap. 2 that the Census of India provides six-fold classification of urban centres. The size class of urban centres varies from the size class consisting of cities, i.e. urban places with 100,000 and more at the top to the size class comprising of the tiny towns, i.e. places having population less than 5000 at the bottom. Researchers sometimes either club them or even sub-categorise them depending upon the emerging situation in the country or in the states. Generally, the bottom three categories are clubbed and designated as small towns. These three categories are towns namely having population 10,000–19,999, 5000–9999 and less than 5000, respectively. The urban places with population in between 20,000 and 49,999 are termed medium towns, whereas large towns consist of population size in the range of 50,000 to 99,999. On the other hand, cities can be divided into million plus and mega cities. Cities with a population of 1 million and more are termed as million plus cities (Bhagat, 2005; Census of India, 1991), while the term mega cities signifies cities at the top of the urban hierarchy and is generally defined as cities either with five, or eight or ten million population depending upon researcher's choice (Kraas & Mertins, 2014). Mega cities are dealt with in Chap. 6, and are defined as a city with 5 million and more population.

The urban population growth rate by size class of urban centres can be estimated based on two approaches namely (i) instantaneous approach and (ii) continuous approach. The instantaneous approach simply considers the population change within the size class category at two points of time. However, during the period between two censuses, several new towns come up and some of the old towns get declassified. The instantaneous approach does not make adjustment for this. Thus, the results of urban population growth rate by size class of towns derived from instantaneous approach are ambiguous. On the other hand, the continuous approach computes urban population growth rate based on population change of only those urban centres, which are common to the two censuses (Census of India, 1991). It is, therefore, an adjusted rate for new towns and also for declassified towns during the inter-censal period. The size class at the later census was taken in the calculation of urban growth rates in this study. This controls the shifts of towns across size classes during the two censuses affecting their growth rates. The later approach is followed in this chapter.

As discussed in Chap. 2, an outgrowth of a statutory urban centre is treated as urban and together with adjoining town (s) is termed as an urban agglomeration (UA). In the calculation of growth rate, researcher can use either the concept of UA or individual town/city. This study has used the concept of individual town/city in the calculation of urban population growth rates based on continuous approach. Thus, population growth figures exclude outgrowths, new towns as well any declassified towns between two census periods. Looking at the dominant role of million plus cities in the urbanisation process of the country, their growth rates are also presented separately in the section below followed by the analysis of population growth rates by size class of urban centres.

## 4.3 Million Plus Cities

The million plus cities defined as cities with population of a million and more merit a special mention because of their large size and economic dominance in India's urbanisation. Million plus cities are also known metropolis literally means a mother city, a meaning no longer holding any significance, though it does mean a dominant or a large city (Dikshit, 2003). According to census 2011, there are 52 million plus cities consisting of 160.7 million urban population and constitute nearly 42.6 per cent urban population of India defined on the basis of the concept of UA (see Table 4.1). Kolkata was the only million plus city at the beginning of twentieth century. Mumbai joined the rank of million plus cities in 1911. Nearly for four decades, there were only two million cities until 1951 when Delhi, Chennai and Hyderabad joined the rank of million plus cities in 1951 increasing the total number of million cities to five. In the decade 1981–91, number of metropolises increased from 12 in 1981 to 23 in 1991 adding 11 more during the decade.

Similarly, during the subsequent decade 1991–2001, the number of metropolises increased to 35 in 2001 from 23 in the 1991 census. The census of 2011 reported total number of million plus UA/cities 52, i.e. an increase of 17 since the 2001 census. It

**Table 4.1** Number and share of population in million plus cities in India, 1901–2011

| Census year | Total UA and towns @ | Number (million plus UA and Towns) | Total urban population (in million) | Population (in million) | Population per UA and Towns | Population per million plus city (UA) (in million) | % to urban population |
|---|---|---|---|---|---|---|---|
| 1901 | 1827 | 1 | 25.8 | 1.51 | 14,152 | 1.51 | 5.84 |
| 1911 | 1815 | 2 | 25.9 | 2.76 | 14,297 | 1.38 | 10.65 |
| 1921 | 1949 | 2 | 28.0 | 3.13 | 14,413 | 1.56 | 11.14 |
| 1931 | 2072 | 2 | 33.4 | 3.41 | 16,150 | 1.70 | 10.18 |
| 1941 | 2250 | 2 | 44.1 | 5.31 | 19,628 | 2.65 | 12.23 |
| 1951 | 2843 | 5 | 62.4 | 11.75 | 21,964 | 2.35 | 18.81 |
| 1961 | 2365 | 7 | 78.9 | 18.10 | 33,377 | 2.58 | 22.93 |
| 1971 | 2590 | 9 | 109.1 | 27.83 | 42,129 | 3.09 | 25.51 |
| 1981 | 3378 | 12 | 157.6 | 42.12 | 46,679 | 3.51 | 26.41 |
| 1991 | 3768 | 23 | 215.7 | 70.66 | 57,264 | 3.07 | 32.54 |
| 2001 | 4378 | 35 | 286.1 | 107.88 | 65,354 | 3.08 | 38.60 |
| 2011 | 6171 | 52 | 377.1 | 160.51 | 61,109 | 3.03 | 42.56 |

*Note* As per provisional figures Chandigarh was a million plus city, but it could not qualify in the final population data released later. *Source* Census of India 2011, A-04: Towns and urban agglomerations classified by population size class with variation between 1901 and 2011, Class I to Class VI; https://censusindia.gov.in/census.website/data/census-tables, accessed on 12 May 2023; Excludes the figures of Jammu & Kashmir where census was not conducted in 1991. Also excludes the figures of Assam where census was not held in 1981. @ This number is based on UA definition and town. When we apply only the definition of town, the total number of towns were 7933 as per 2011 census.

is also interesting to note that the average population of 3.5 million per metropolis, i.e. maximum reached in 1981 remains almost unchanged in spite of a significant increase in the number of new metropolises during the subsequent decades (Bhagat, 2005). Thus, it is not simply the concentration of population but the geographical expansion of million plus cities that is significantly shaping India's urbanisation. The geographical expansion and increase in number of million plus cities led to the rise from five per cent of the urban population living in million cities in 1901 to 20 per cent in 1951. The share of urban population in million plus cities further increased to about one-fourth in the 1970s and to more than two-fifths of total urban population in the 2000s (see Table 4.1). It is also evident from Table 4.1 that the number of urban centres (UA and towns) have increased from 1827 in 1901 to 6171—almost three times increase compared to 15 times increase in total urban population. On the other hand, million plus cities show a pole-vaulting increase of 106 times in their population in 2011 compared to 1901.

A huge increase in the concentration of urban population in million plus cities is indicative of the fact that urbanisation process in India shows a form of concentrated urbanisation. Interestingly the concentrated form of urbanisation is also accompanied by extensive and differential forms of urbanisation as well. The extensive form of urbanisation reflects in the process of emergence of large number of census towns during 2001–2011. Further, the differential urbanisation is also visible in terms of growth rates of core and peripheral areas of million plus cities (Bhagat, 2004, 2018). Such pattern of population growth is not a new phenomenon and has been observed in earlier decades as well which Brush (1962) described as a centrifugal force shaping the process of urbanisation and urban sprawl.

In general, the nature of migration and population growth in the million plus cities UAs has changed moving from the core to peripheral areas. The core area is defined as main Municipal Corporation areas showing lower population growth and lower migration rates compared to peripheral areas outside the main municipal areas but within the UA in several such cities. A detailed analysis of migration data by mega city UA is presented and discussed in Chap. 6.

Here it is important to mention that one of the important features of India's urbanisation emerging from the 2011 census is not only faster urbanisation, but also the faster urbanisation has been accompanied by extensive urbanisation showing geographical expansion through the emergence of new towns, and also manifesting in the dominant share of urban population in cities in general and million plus cities in particular.

The trend in urbanisation in India in terms of three identified features namely extensive, differential and concentrated is not an exception. Brenner and Schmid (2015) think that all three aspects are important features of planetary urbanisation triggered by global capitalism. Some of the features of India's urbanisation highlighting the pattern of concentrated and differential urbanisation are presented below.

## 4.4 Concentrated and Differential Urbanisation: Analysis by Size Class of Cities and Towns

There were 6171 UA and towns consisting of 474 UA and 5697 towns as per the 2011 census. As discussed above out of total UA and towns, there were 52 million plus cities which comprised 42.6 per cent of total urban population of India in 2011. Figure 4.1 shows that about 8 per cent of all UA and towns belong to Class I urban centres (468 including 52 million plus cities) which constitute about 70 per cent of India's urban population. On the other hand, small towns (less than 20 thousand) comprised of about 63 per cent of all UA and towns constituted about 10 per cent of total urban population in 2011. This shows huge concentration of population in Class I (100 thousand and more) cities in general and million plus cities in particular. The nature of urban population concentration is also depicted in Lorenz curve presented in Fig. 4.2, and the Ginni co-efficient was urban population concentration is 0.77 in the scale of 0 to 1. The value of zero represents that urban population is evenly distributed among size class of urban centres, while 1 pertains to entire population located in a particular size class of urban centres. A Ginni co-efficient 0.77 stands for

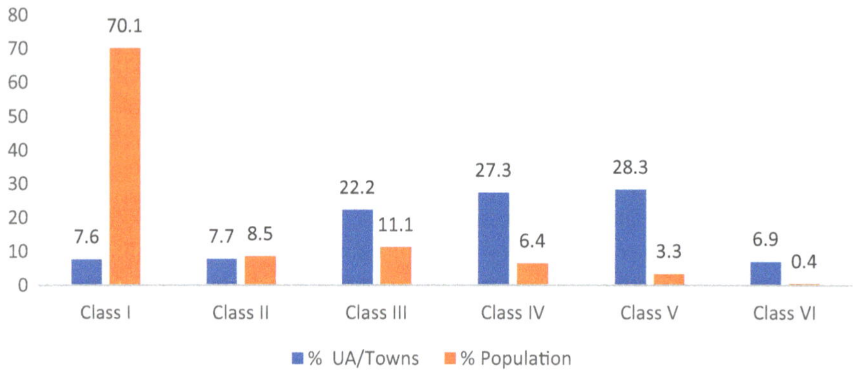


**Fig. 4.1** Distribution of number of UA/Towns and share of population by size class of urban centres, India, 2011 (*Source* Same as in Table 4.1)

a high level of urban population concentration as evident from the fact that two-fifths of urban population falls under the category of million plus cities.

Very interestingly Fig. 4.1 also shows that tiny towns (Class VI) with population less than 5000 constitute 0.4 per cent population of all urban population and comprise of about 7 per cent of all UAs and towns. Their number was 424 as per the 2011 census. A question arises why have towns with less than 5 thousand whereas the urban definition of census stipulates 5000 population as a minimum criterion? Reasons are both technical as well as discretionary criterion used in declaring a settlement as urban. As per instruction of the Census of India the settlements with 4000 in the previous census were identified and assumed that these will qualify to have 5000 population in the subsequent census after a period of 10 years' time. However, in actual practice many of them do not reach the mark of 5000, although already classified as urban based on criteria like size, density and male main non-agricultural workforce. Some of them with more than 5000 in 2001 even declined to below 5000 due to out-migration during 2001–2011. The number of such towns was 424 consisting of total of 1.6 million in 2011. Interestingly most of these towns also fall under the category of census towns. Another important reason pertains to the discretionary power of the Directors of Census Operations in the States and Union Territories who in consultation with the state governments/Union Territories and the Census Commissioner of India are allowed to include some places having distinct urban characteristics as urban even if such places did not strictly satisfy all the criteria mentioned in the non-statutory part of the urban definition (Census of India, 1991).

While population concentration has been rising in big cities, at the same time share of population has been declining in towns (less than 100 thousand) in general and small towns in particular in spite of their rising number. Table 4.2 shows urban population distribution by size class of cities and towns since 1901. It may be noted that the share of small towns (less than 20 thousand) was about half of the urban population in 1901 which declined just one-tenth by 2011. Same characteristics are found in other categories of towns as well but with lesser intensity.

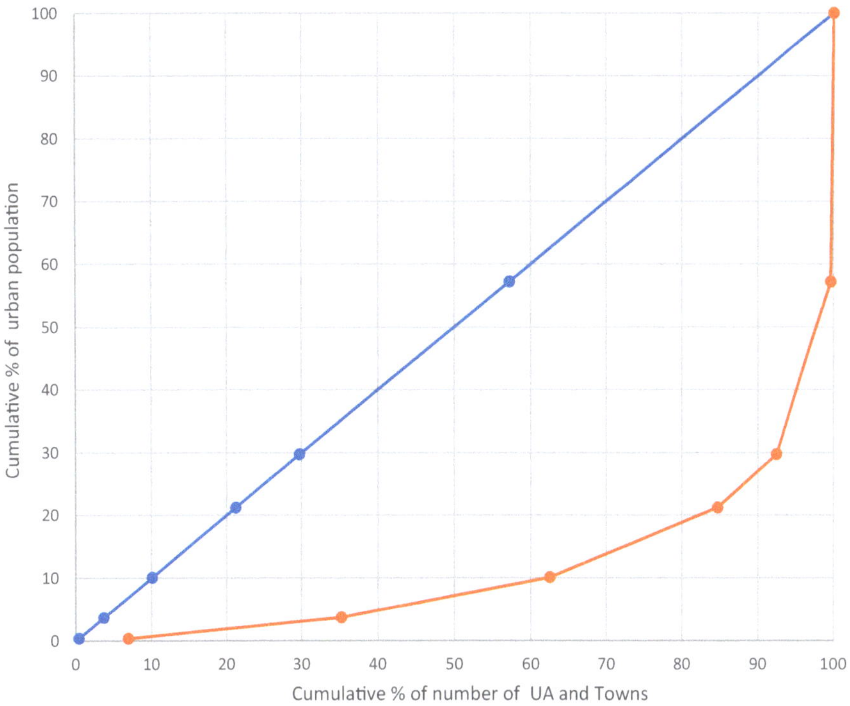


**Fig. 4.2** Lorez curve showing concentration of urban population by cumulative share of number of UA and towns, India, 2011 (*Note* Ginni co-efficient is 0.77, which shows urban population concentration in 0 to 1 scale)

**Table 4.2** Urban population by size class of cities and towns, India, 1901–2011

| Census year | Million cities (one million & above) | Cities (100 thousand to one million) | Large towns (50 to 100 thousand) | Medium towns (20 to 50 thousand) | Small towns (less than 20 thousand) |
|---|---|---|---|---|---|
| 1901 | 5.86 | 20.11 | 11.29 | 15.64 | 47.10 |
| 1911 | 10.89 | 16.74 | 10.51 | 16.40 | 45.46 |
| 1921 | 11.30 | 18.40 | 10.39 | 15.92 | 43.99 |
| 1931 | 10.34 | 20.86 | 11.65 | 16.80 | 40.35 |
| 1941 | 12.19 | 26.04 | 11.42 | 16.35 | 34.00 |
| 1951 | 19.07 | 25.57 | 9.96 | 15.72 | 29.69 |
| 1961 | 23.34 | 28.08 | 11.23 | 16.94 | 20.41 |
| 1971 | 26.02 | 31.22 | 10.92 | 16.01 | 15.83 |
| 1981 | 26.93 | 33.49 | 11.63 | 14.33 | 13.62 |
| 1991 | 33.18 | 32.01 | 10.95 | 13.19 | 10.66 |
| 2001 | 37.80 | 30.78 | 9.73 | 12.29 | 9.36 |
| 2011 | 42.56 | 27.63 | 8.53 | 11.09 | 10.15 |

*Source* Same as in Table 4.1

However, the concentration of population in big cities is not accompanied by primacy of a single city in India's urban system and urban hierarchy. This is evident in the fact that the evolving urban hierarchy follows a rank-size rule (Fig. 4.3). The regional structure of India's urbanisation plays an important role which is unified by the mega cities like Mumbai, Delhi, Kolkata and Chennai located in different parts like west, north, east and south India. The rank-size rule of urban system has been expounded by G.K. Zipf (1941) to explain the bio-social evolution of a nation state which explains the underlying mechanism of the relationship between rank of the city and its size in an evolving capitalist economy. Zipf also identified two types of forces namely forces of unification and forces of diversifying in the evolution of urban hierarchy of an urban system defined on the basis of rank-size rule. It is also based on an idea of inequality which follows pareto distribution reflected in 80/20 pareto principle. It means that 80 per cent of outputs come from 20 per cent causes. Following this principle 80 per cent of urban population must reside in 20 per cent cities and towns. This is very much evident from the Lorenz Curve presented in Fig. 4.2.

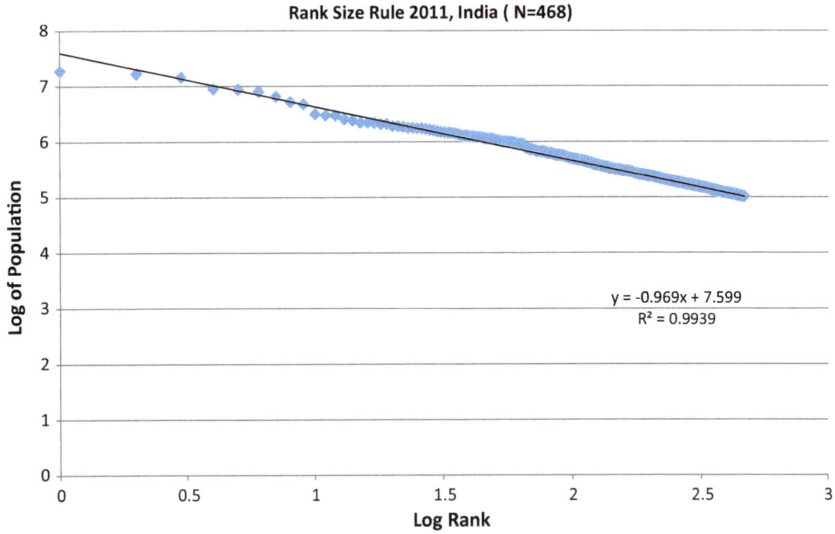


**Fig. 4.3** Urban hierarchy and rank-size rule, India, 2011

## 4.5 Urban Population Growth by Size Class of Towns and Cities

It is generally believed that bigger cities are growing faster than medium and small urban centres. To know that it is pertinent to calculate population growth rate by size class of cities and town based on continuous approach. Table 4.3 presents the adjusted decadal growth by size class of cities and towns for the period 1981–2011. The size class has been taken for the latter census and population of common cities and towns during the two-census period has been used to calculate decadal growth rate. As the adjusted rates are based on common towns during the period and excludes the new and declassified towns, the extent of growth is largely contributed by natural growth and migration and also to some extent by the changes in the boundaries of the towns/cities. The adjusted rates are lower than the unadjusted rates for the decades 1981 to 2011 at the country level (Bhagat, 2005).

The adjusted decadal urban population growth rate for India was 21.7 per cent during the decade 2001–2011 (see Table 4.3). Compared to this the decadal urban population growth rate was observed at 31.8 per cent during the same period at the country level. The adjusted growth rates show that the decadal urban population growth rate has declined from about 28 per cent during 1991–2001 to 21 per cent during 2001–2011 (see Table 4.3). The decline in the adjusted decadal urban population growth rate is also reflected across all size class of urban centres. This is expected as overall population growth rate has been declining. Notwithstanding, it is very important to observe from Table 4.3 that cities and small towns have grown at higher rates compared to large and medium towns during 2001–2011, while it was not so pronounced in the preceding decade during 1991–2001. In an earlier study of urban population growth rate by size class of cities and towns based on continuous approach prepared by Jain, Ghosh and Kim as an occasional paper of Census of India showed that the population growth of rates by size class of urban centres do not reveal any significant differentials during the decade 1961 to 1991. This study concluded that neither city has registered a rapid pace of population growth, nor the

**Table 4.3** Decadal per cent change by size class of cities and towns, India 1991–2011, continuous approach[a]

| Size class | 1991–2001 | 2001–2011 |
|---|---|---|
| Cities[b] (>100,000) | 28.4 | 23.7 |
| Large Towns (50,000–100,000) | 26.1 | 11.3 |
| Medium Towns (20,000–50,000) | 24.5 | 16.2 |
| Small Towns (<20,000) | 28.3 | 26.5 |
| Total | 27.5 | 21.7 |

[b] Cities are inclusive of million plus cities. *Note* [a] Decadal per cent change exclude the new towns as well declassified towns and are based on common towns during the decade. *Source* Same as in Table 4.1

small towns have shown any stagnation or a decline in India during last thirty years period (Census of India, 1991). Similar results are also vindicated from the results presented in Table 4.3 based on subsequent decades.

It may be observed from Table 4.3 that cities have shown an average decadal population growth rate of 28.4 per cent compared an average decadal population growth rate of 27.5 per cent among small towns during 1991–2001, while small towns exhibit higher decadal growth rate of 26.5 per cent compared to 23.7 per cent among cities during 2001–2011. It may be construed as an important sign of India's urbanisation history wherein not only we have seen a huge contribution of census towns mostly belonging to the category of small towns (less than 20 thousand) in urban growth but also the existing small towns have also shown higher growth rate compared to any size category of cities and towns during the 2000s. It means that a part of the rural to urban migration is directed to these small towns as well. On the other hand, the decadal population growth in large and medium towns in the range of from 20 thousand to less than 100 thousand have declined faster showing that the people of these towns migrating to cities, million plus and mega cities. The trend although started during the 1990s has got strengthened during the 2000s. Migration data also show that there has been a large increase in urban to urban migration during 2001–2011. The emerging trend of urban population growth by size class of urban centres is indicative of urban transition not only unleashed by rural to urban migration but also influenced by urban to urban migration as well. This led to urban population growth rate higher at both ends of the urban hierarchy. Preston (1979: 201) also observed that the general relationship between urban size and population growth rates in developing regions is U-shaped. It means that the largest and the smaller urban centres grow faster.

This is consistent with forces of privatisation and liberalisation of the economy unleashed during the 1990s indicating the trend that migration to big cities is more selective compared to small urban centres. Further, a person with some education and skill is more equipped to survive in cities in general and big cities in particular. It is in this context that urban to urban migration assumes significance. On the other hand, rural to urban migration of short distance nature is mainly directed to small and medium towns attracting various strata of rural society like landed class who like to have a second home in the nearby town, and petty business folks and entrepreneur classes and students flocking to nearby towns for getting higher education in addition to activities like rickshaw pulling and working in the transportation sector and in trade and commercial activities. In addition, as the cost of living in big cities has been rising along with increasing precarity of jobs predominantly available in the informal sector with the shrinkage of the organised sectors (Kundu, 1997; Planning Commission, 2001) has a dampening effect on migration to big cities contrary to the popular perception of rural to urban migration. In such a situation migration to the nearby towns is an alternative possibility left to the rural people (Bhagat, 2005). It may be clarified that such migration does not include seasonal, temporary and circulatory nature because it does not involve change in the usual place of residence, and as such does not qualify to be counted as migrant at the place of destination as per the standard definition of migration.

The population growth by size class categories also shows that there are no consistent variations at the state level (see Table 4.4). The pattern shown at India level is also evident in most of the states. In general, in most of the states, cities and small towns are showing higher population growth rates compared to large and medium towns. In states of Gujarat, Karnataka, Haryana and Andhra Pradesh cities have shown higher population growth rate and the same is also true for small towns as well. On the other hand, in Kerala, Tamil Nadu, Goa, Maharashtra and West Bengal cities have shown lower population growth and the same is also true for their respective small towns as well (see Table 4.4).

A scatter plot of population growth of cities and small towns fitting the trendline also shows a positive relationship between the two size class categories (see Fig. 4.4).

**Table 4.4** Decadal per cent change in population by size class of towns/cities, India and states, 2001–11 (continuous approach)[c]

| States | Cities (>100,000) | Large towns (100,000–50,000) | Medium towns (50,000–20,000) | Small towns (<20,000) | Total |
|---|---|---|---|---|---|
| Andhra Pradesh | 33.3 | 12.7 | 18.2 | 21.4 | 29.2 |
| Assam | 14.7 | −0.7 | 4.9 | 44.9 | 17.7 |
| Bihar | 19.6 | 28.0 | 21.0 | 14.6 | 20.9 |
| Chhattisgarh | 29.1 | 20.7 | 25.8 | 19.5 | 26.7 |
| Delhi | 12.4 | 33.0 | 43.6 | 42.3 | 16.0 |
| Goa | | −3.9 | 1.3 | 10.8 | 4.1 |
| Gujarat | 36.2 | 6.0 | 17.0 | 69.3 | 33.3 |
| Haryana | 41.0 | 11.6 | 20.0 | 29.8 | 36.4 |
| Jharkhand | 19.7 | 19.8 | −2.0 | 80.0 | 27.0 |
| Karnataka | 45.2 | 12.6 | 13.9 | 13.9 | 37.6 |
| Kerala | 0.1 | 4.6 | 5.6 | 6.0 | 2.5 |
| Madhya Pradesh | 21.0 | 14.2 | 17.2 | 14.6 | 18.7 |
| Maharashtra | 15.6 | 13.8 | 14.4 | 15.9 | 15.3 |
| Odisha | 26.6 | 12.9 | 17.2 | 11.4 | 20.7 |
| Punjab | 17.3 | 14.1 | 21.1 | 15.5 | 17.1 |
| Rajasthan | 22.2 | 19.1 | 15.1 | 14.3 | 19.9 |
| Tamil Nadu | 10.8 | 3.8 | 22.8 | 22.8 | 15.2 |
| Uttar Pradesh | 23.5 | 10.6 | 16.9 | 35.4 | 23.0 |
| Uttarakhand | 26.2 | 11.0 | 16.3 | 16.3 | 19.7 |
| West Bengal | 6.2 | 10.3 | 12.4 | 19.3 | 8.3 |
| India | 23.7 | 11.3 | 16.2 | 26.5 | 21.7 |

[c] Decadal per cent change excludes the new towns as well declassified towns and is based on common towns during the decade. *Source* Same as in Table 4.1

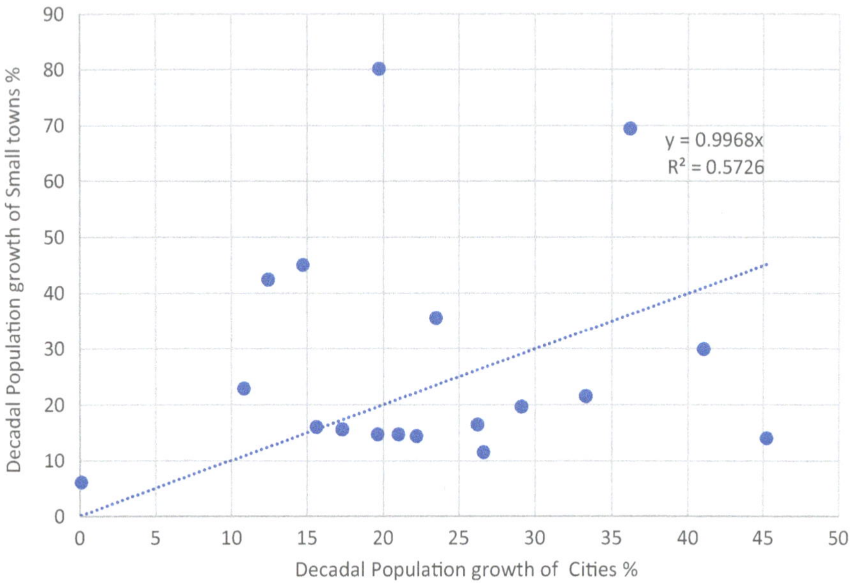


**Fig. 4.4** Relationship between the population growth of cities and the population growth of small towns at state level, 2001–2011

We conclude that the country level pattern of population growth by size class of cities and towns characterised as closer to the U pattern described earlier is also by and large evident at the state level as well. It means that urban centres by size class are growing relatively higher at both the high and low ends and there is a dip in the population growth in the middle size categories. This shows rural to urban migration is also directed to the small urban places which are located in the proximity of rural areas and the rising nature of urban to urban migration from the medium and small towns to the big cities. Thus, the emerging pattern of urbanisation needs to take into account the urban demographic change by the size class of urban centres. Thus, in order to know whether urban size of cities and towns has any impact on their growth rates, the size of cities and towns as an independent variable is regressed on their demographic growth rates as a dependent variable during 2001–2011. The natural log of the size of each city/town has been taken into consideration in the analysis. The regression analysis reveals that size of the town/city is negatively associated with growth rates, i.e. higher is the size lower is the population growth rate, however, not significant. A similar finding was also observed in other studies as well (Bhagat, 2004; Shaban et al., 2020).

A negative relationship between the size of the towns/cities and their respective population growth rates generally supports a theory of urban transition premised on the fact that the continuous increase in size can not be sustained in the long run. A town or a city within its fixed territorial limits is likely to experience decline as size increases. It is evident as cities grow their economic base and activities also expand leading to increased advantage arising from agglomeration economies. However, it cannot be sustained ever as an increase in the size and density after a threshold also increases diseconomies due to crowding, congestion, pollution, scarcity of land, its rising prices, etc., eventually leading to the decline in the population growth rate of an urban centre when the areal expansion of the urban centre is kept constant. On the other hand, areal expansion of a city/town redefines its boundaries and also its size. Thus, the size of an urban centre is a dynamic and evolving concept, and the city and urban planning should respond to the emerging needs and challenges of urban transition (Mills & Becker, 1986). Cities should be allowed to grow naturally in order to reap the benefits of its growth momentum. In fact, the optimality of city size is elusive and each city could find its own in due course of time (Bhagat, 2005). However, it is pertinent how municipal boundaries are drawn for administrative efficiency and developmental planning of urban areas. This is to be understood as a dynamic process as differential population growth rates by size class of urban centres continue to manifest in varied spatial forms of concentrated and extensive urbanisation. Municipal administration, governance and planning must align and adapt to such changes to promote urbanisation as a vehicle of sustainable development.

## 4.6 Conclusion

India has been increasingly urbanised since the second quarter of the twentieth century; however, slowed down after a peak in the 1970s. A large number of the emergence of the census towns during the 2000s apparently shows that the declining tempo of urbanisation has been reversed. However, we can better understand the process of urban transition when trying to analyse the population growth rates by size class of urban centres. Our analysis shows a differential population growth by size class of urban centres arising from the process of urban transition. Although cities including million and mega cities show higher population growth rates, we also find that small towns are no less behind. In general, the relationship between the size of the urban centres and population growth shows a negative relationship reflecting that smaller urban centres have higher potential for population growth in the course of urban transition. However, concentrated urbanisation continues to be a characteristic feature of India's urbanisation evolved over a century and a half accompanied by extensive urbanisation evident in the emergence of large number of census towns during the 2000s in general, and also contributing to higher population growth in the peripheral areas of million and mega city urban agglomerations in particular along with urban sprawl. The relationship between the concentrated, differential and extensive urbanisation and their varied forms would be an important framework for capturing India's urban transition in the future.

**Acknowledgements** An earlier version of this chapter entitled 'Urban Growth by City and Town Size in India', by the first author was presented in the annual meeting of Population Association of America to be held at Philadelphia, USA, March 31–April 2, 2005, https://paa2005.populationassociation.org/papers/51133.

# Note

1. The concept of primate city refers to the largest city of a country, often the capital city which may be several times larger than the second-largest city. Mark Jefferson (1939) proposed this concept for the first time in an article entitled 'The law of the primate city'. The idea was later developed and elaborated by Zipf (1941) in a statistical regularity known as Rank-Size Rule.

# References

Bhagat, R. B. (2018). *Urbanisation in India: Trend, Pattern and Policy Issues* (Working Paper No. 17). International Institute for Population Sciences. https://www.iipsindia.ac.in/sites/default/files/IIPS_Working_Paper_No_17.pdf. Accessed on October 24th, 2023.

Bhagat, R. B. (2004). Dynamics of Urban Population Growth by Size Class of Town and Cities in India. *Demography India, 33*(1), 47–60.

Bhagat, R. B. (2005, March 31–April 2). *Urban Growth by City and Town Size in India*. Paper to be Presented in the Annual Meeting of Population Association of America to be Held at Philadelphia, USA. https://paa2005.populationassociation.org/papers/51133

Brenner, N., & Schmid, C. (2015). Towards a New Epistemology of Urban? *City, 19*(2&3), 151–182.

Brush, J. (1962). The Morphology of Indian Cities. In R. Turner (Ed.), *India's Urban Future*. University of California Press.

Census of India. (1991). Emerging Trends of Urbanisation in India, Occasional Paper No. 1 of 1993, Prepared by Jain, M. K., Ghosh, M. & Kim, W. B., Office of the Registrar General, and Census Commissioner, India, New Delhi.

Champion, T. (2001). Urbanization, Suburbanization, Counterurbanization and Reurbanization. In R. Paddison (Ed.), *Handbook of Urban Studies* (pp. 143–159). Sage Publications.

Dikshit, K. R. (2003, January 28–29). *Metropolitan Cities: Issues of Development*. Paper Presented in Regional Seminar on Metropolitan Cities in India: Issues of Development, Held at Department of Geography, University of Mumbai.

Geyer, H., & Kontuly, T. (1993). A Theoretical Foundation for the Concept of Differential Urbanisation. *International Regional Science Review, 17*(2), 157–177.

Glaeser, E. (2010). *Agglomeration Economics* (pp. 1–14). Chicago University Press.

Jefferson, M. (1939). The Law of the Primate City. *Geographical Review, 29*, 226–232.

Kraas, F., & Mertins, G. (2014). Megacities and Global Change. In F. Kraas, S. Aggarwal, & G. Mertins (Eds.), *Megacities: Our Global Urban Future* (pp. 1–6). Springer.

Kundu, A. (1997). Trends and Structure of Employment in the 1990s: Implication for Urban Growth. *Economic and Political Weekly, 32*(4), 1399–1405.

Mills, E. S., & Becker, C. M. (1986). *Studies in Indian Urban Development*. Oxford University Press.

Pacione, M. (2001). *Urban Geography: A Global Perspective*. Routledge.

Planning Commission, Govt. of India. (2001). *Draft Approach Paper to the Tenth Five-Year Plan (2002–2007)*. Planning Commission, New Delhi.

Preston, S. H. (1979). Urban Growth in Developing Countries: A Demographic Reappraisal. *Population and Development Review, 6*, 195–215.

Shaban, A., Kourtit, K., & Nijkamp, P. (2020). India's Urban System: Sustainability and Imbalanced Growth of Cities. *Sustainability*, *12*(7), 1–20. www.mdpi.com/journal/sustainability. Accessed on March 26, 2023.
Zipf, G. K. (1941). *National Unity and Disunity: The Nation as a Bio-social Organism.* Principia Press.

# Chapter 5
# Rural to Urban Migration: Trends, Patterns and Policy Issues

## 5.1 Introduction

The history of migration began some 200 thousand years ago when *Homo sapiens* left Africa's Omo Valley creating the earliest days of human history and civilisation producing distinct cultures, languages and ethnicities in different parts of the world (World Bank, 2023: 1). India too has a long history of migration which has shaped its social history, culture and pattern of development. In pre-colonial times, the reasons for the circulation of populations were mainly for religious and trade purposes (McNeill, 1984: 9). Migration, on account of military movements, also played an important role. People also travelled in search of pastures with their cattle. Nomadic migration, even for short distances, was an important feature outside the Gangetic Valley. This practice is still found in some parts of Rajasthan and Madhya Pradesh located in central India and is one of the earliest forms of circulation in the history of human migration (Hutton, 1986: 61).

Historically migration is closely related to urbanisation. In India, urbanisation is not a new phenomenon but is an important feature of India's ancient and medieval history. In ancient India, there were many great urban centres. For example, around 300 BC, it is estimated that *Patliputra* (the present Patna) had 270 thousand people, *Mathura* had 60 thousand, 48 thousand people lived in *Vidisha*, 40 thousand each in *Vaishali* and *Kaushambi*, and 38 thousand lived in *Ujjain* (Sharma, 2005). At the death of the Emperor Akbar, India had a population of 100 million in medieval times and Agra was perhaps one of the largest cities in the world. The Moghul emperors used to set out with an army and camp in several places during their military expeditions. This not only included soldiers but also a vast number of people who were manning the hundreds of accompanying *bazaars* (markets). Such military expeditions were themselves forming temporary cities which were constantly on the march (Nehru, 1965).

When the British came to India, the old feudal order was breaking up. The fall of the Moghul Empire produced political chaos and disorder in many parts of India. But



R. B. Bhagat and M. I. Hassan, *Urbanisation and Urban Policies in India*, Exploring Urban Change in South Asia, https://doi.org/10.1007/978-981-96-3181-0_5

even so, India in the eighteenth century was a great manufacturing as well as a great agricultural country, and Indian hand-looms were supplying the markets of Asia and Europe. However, in the face of machine-made garments produced in the mills of Lancashire and Manchester in Britain, the Indian fabric (both silk and cotton) and allied industries (carding, dying and printing) could not survive. The British also imposed transit duties (*octri*) on the internal movement of goods within India. This further crippled the Indian hand-based Industries. A vast number of Indian weavers and artisans became jobless and unemployed. India experienced a massive deindustrialisation between 1757 and 1857 when the East India Company of Britain ruled the country. Most of the weavers and artisans lived in towns and cities. As a consequence of deindustrialisation, a large number of the artisan class and their dependent families migrated to the countryside. Thus, India during the second half of the eighteenth century to the first half of the nineteenth century experienced urban to rural migration unlike many European countries that underwent massive rural to urban migration during the same time. As a result, cities and towns declined and languished and people became more dependent on agriculture (Nehru, 1965: 433). The economic hardship and economic disturbances were other important reasons for the migration of people in the late eighteenth and early nineteenth century. However, there are indications that mobility from the countryside to town was limited through the operation of guild or guild-like restrictions on employment in urban trades. Furthermore, the structure of production and consumption patterns that guided urban output placed a major barrier to the growth of a rural–urban migratory process on account of the limited access to unskilled and casual labourers (Commander, 1989).

After the great mutiny of soldiers in 1857, the British government took direct charge of India. In the post-1857 period, the development of plantation agriculture, including tea, rubber, jute, cotton, indigo and mining and quarrying was an important factor in internal migration. The two areas of plantation agriculture that emerged—namely Assam in the north-east and Travancore and Mysore in south India—were dependent on migrant labour. The labour was indentured in nature. The migrants were generally placed under a five-year indenture containing penal provisions for breach of the agreement. Since plantations were located in the remote wilderness with very little communication with the rest of India, the workers were virtual prisoners for the duration of their contact. On the other hand, the growth of plantation agriculture and mining and quarrying led to the development of railways and ports, which established trade relations with the colonies of the British Empire and with Great Britain. This reorganised the space relations, pattern of migration and circulation which was mostly directed towards new foci in the western, southern and eastern India. Ports like Bombay, Calcutta and Madras were established. These new cities did not exist in ancient or medieval times but came to exert a great influence on the pattern of circulation and migration in the Indian subcontinent in colonial and post-colonial India. However, only with the beginning of the census in the late nineteenth century was it possible to study the trends and patterns of migration in general and rural to urban migration in particular.

## 5.2 Definitions and Migration Data

While migration has been a part of human history since the beginning, migration data has been available on a required scale since the beginning of the Imperial Census in India in the latter half of the nineteenth century. The first synchronous census was conducted in 1881 and a definitive information on migration based on place of birth (POB) only became available in the 1881 census and has continued in all censuses conducted to date. However, the POB data on migration provides only lifetime migration. As it is not known to which migration move it refers and when it occurred, it has limited use in the study of migration patterns and its characteristics. POB migration data, however, does not cover return migration. Realising the limitation of POB data on migration, the Census of India has since 1971 also provided migration data on the basis of place of last residence (POLR). The POLR data is also classified by duration of residence at the place of enumeration. Unlike POB, which provides information on lifetime migration, POLR gives information on the last move of migrants and also captures return migration. As per census procedure, if the POB/POLR is different from the place of enumeration (POE), the person is classified as migrant. Villages and towns, whose boundaries are administratively defined, are the lowest units for determining POB/POLR. Villages are defined as revenue villages which comprise areas occupying human habitation as well as areas of other land uses. It is also possible that some revenue villages may be uninhabited. Any change of residence beyond the village or town/city boundary qualifies a person to be classified as a migrant. At the town level, migration data are not available for all urban centres but mostly for Class I cities with a population of more than 100,000. However, it is important to mention that migration data, while not available for administrative towns/cities, exist at the urban agglomeration (UA) level. An urban agglomeration is a continuous urban area consisting of a town or city and its adjoining outgrowths (OGs), or two or more physically contiguous towns/cities together with or without their adjacent outgrowths. As POLR and POE are classified by rural and urban areas, it is possible from census data to study the various streams of migration such as rural to rural, rural to urban, urban to urban and urban to rural. It should be noted that, while the rural–urban status of POE is classified based on the urban definition followed, the rural and urban status of POLR is collected at the household level during census enumeration. As a result, a certain number of rural and urban migrants are unclassifiable (13% of urban migrants are unclassifiable in the 2001 census compared to 10 per cent in the 2011 census) in the streams of migration. Further data on migration are provided, for instance, a change in residence elsewhere in the district (intra-district); from one district to another within the state (inter-district); from one state to another (inter-state) and from another country. The duration of migration is also available through POLR criterion which provides the duration of stay at the present place since the last move of the migrant. Such data in censuses appear to be badly affected as the unstated duration has kept on rising particularly that relating to males. About one-fifth of urban male migrants did not report duration in the 2001 census, and the same proportion continued even in the 2011 census. Compared to urban male migrants, the duration not reported for urban female migrants was one-tenth only in both censuses.

Districts are the lowest unit for which migration data are available. There were 593 districts in the 2001 census which increased to 640 in the 2011 census. An increase in the number of districts could greatly affect intra-district and inter-district migration data and it is, therefore, advisable to compare the within and the between states migration which largely remains unaffected by the number of districts. In censuses, immigrants are also enumerated but the Indian census, being de facto, does not provide information on the emigration of Indians. It is, consequently, not possible to study net international migration from census data though net inter-state migration in India can be examined. Since the 1981 census, the reasons for migration have been added to the census schedule. Apart from the census, the National Sample Survey Office (NSSO)—a wing of the Ministry of Planning and Programme Implementation—also now includes a question on migration based on POLR in its employment and unemployment surveys. In NSSO surveys, the POLR is defined as a place (village or town) where the migrant has stayed continuously for a period of six months or more before moving to the place of enumeration. In contrast, censuses do not limit the duration of residence in defining the POLR. In both censuses and NSSO, there is no limit of duration at the POE to qualify as a migrant. Migrants are identified as such if the POLR is different from the POE.

In most of the demographic studies, migration is analysed in terms of stocks which denotes that the study is concerned with the number of migrants. On the other hand, a migrant might be making many moves in his or her life cycle but national datasets like Census and NSSO surveys are not able to capture all these moves but provide data only on the last move based on POLR. All those who move within the village and municipal boundaries are not considered migrants. In some cases, where municipal areas are very large, intra-urban migration could be significant.

It is alleged that Census and NSSO data are not able to fully capture the 'short-term migrants' on the basis of POB and POLR criteria as short-term migration is seasonal, temporary and circular in nature. NSSO has tried to capture short-term/temporary migrants using a different question. For example, in the 64th Round (2007–08), each household was asked whether any member had gone away for employment purposes for more than one month but less than six months during the previous year.

The same definition was also employed in the 55th Round (1999–2000) except that the duration was for more than two months (NSSO, 2010).[1] In subsequent PLFS (Periodic Labour Force Survey) and Multiple Indicator Survey, the question on short-term migration was not included. Notwithstanding, the Census and NSSO data are enormously helpful in the study of migration, reflecting changes in the Indian economy and society.

## 5.3 Urban Migration Trends

Kingsley Davis (1951) in his pioneer work on India argued that Indians were less mobile. This conclusion was based on inter-provincial/state migration which stood at 3.6 per cent in 1931 in India compared to 23 per cent in the United States in 1940. Davis attributed this fact to the prevalence of the caste system, joint families, traditional values, the diversity of language and culture, the lack of education and the predominance of agriculture and semi-feudal land relations in India. However, the fact remains that Indian migration is predominantly a 'within state' phenomenon. According to the 2001 census shows inter-state migration based on POLR is only about 4 per cent (41 million) of India's population compared to 26 per cent within state migration (268 million) (Bhagat, 2010). Thus, taking into account the entire mobility including within and between state migrations, the mobility of Indian population stands close to 30 per cent—much higher than what Davis believed. It is also startling to see that about 30 per cent of India's population is internally mobile, a figure which has remained remarkably stable over the last several decades and is also consistent with NSS data. The 1971 census shows 29 per cent internal mobility compared to 30 per cent in 2001 and the NSS data for the year 2007–08 shows about 29 per cent of India's population are internally mobile. As per 2011 census the share of migrants rose to 37 per cent in India. The increase was largely due to within state migration.

India is a federal country. At the time of the 2011 census, it comprised 28 states and 7 union territories. A new state of *Telangana* was created in 2014 from the state of Andhra Pradesh increasing the number to 29. At the state level, there is a marked variation in the level of urbanisation. States like Bihar, Assam and Himachal Pradesh are at the bottom of the ranking where the urbanisation level is less than 15 per cent compared to the states of Tamil Nadu, Punjab, Maharashtra and Gujarat where the urbanisation level, ranging from 37 per cent in Punjab to 48 per cent in Tamil Nadu, is much higher than the national average. These highly urbanised states show a higher level of per capita income and also a higher level of in-migration. The National Capital Territory of Delhi is another important state that stands out alone with huge in-migration. The megacity of Delhi along with Kolkata, Mumbai and Chennai (Madras), which emerged as port cities during the colonial rule, shaped the regional pattern of urbanisation and the flow of migration. Any understanding of the trend and pattern of migration in India is basically rooted in the emerging pattern of urbanisation and the urban system that evolved during the colonial rule and was reinforced in independent India (Bhagat, 2012a; Raza & Habeeb, 1976).

Table 5.1 shows that about 35 per cent of India's urban population is constituted of migrants according to NSS Survey 2007–08. The subsequent rounds of PLFS and Multi Indicator Survey also confirm the magnitude of migration intensity in urban areas. On the other hand, the census gives a higher proportion of migrants as high as 46 per cent in the 2011 census. The difference between the two sources lies in defining the POLR as mentioned earlier, and also some of the categories like moved after birth unlike NSS Survey.

**Table 5.1** Percentage of migrants* in urban areas, India 1981–2021

| Census/NSS Years | Census | | | NSS | | |
|---|---|---|---|---|---|---|
| | Male | Female | Total | Male | Female | Total |
| 1981/1983 | 33.2 | 40.8 | 36.8 | 27.0 | 36.6 | 31.6 |
| 1991/1993 | 26.3 | 36.2 | 31.0 | 23.9 | 38.2 | 30.7 |
| 2001/1999–2000 | 32.0 | 39.4 | 35.5 | 25.7 | 41.8 | 33.4 |
| NA/2007–2008 | NA | NA | NA | 25.9 | 45.6 | 35.4 |
| 2011/NA | 40.8 | 52.10 | 46.2 | NA | NA | NA |
| PLFS 2020–21 | NA | NA | NA | 22.5 | 47.8 | 34.9 |
| Multiple Indicator Survey 2020–2021 | NA | NA | NA | 22.3 | 47.8 | 34.6 |

*Source* Census of India 1981 to 2001, D-2 migration Table; migration rates exclude Assam for 1981 and Jammu and Kashmir for 1991 where the census was not conducted. D-2 Table is based on the criterion of Place of Last Residence (POLR)

* All tables in this chapter refer to permanent/semi-permanent migrants defined as considering a change in the usual/habitual place of residence

NSS figures are taken from NSSO (2010: 23), NSO (2022) Periodic Labour Force Survey (PLFS), 2020–21; NSSO (2023) Multiple Indicator Survey in India, 2020–2021. NSS figures are based on the criterion of Place of Last Residence. As PLFS and Multiple Indicator Survey are conducted during the same reference year, only PLFS data are presented in subsequent tables

There was a decline in the rate of migration to urban areas between the Census/NSS years of 1981–1983 and 1991–1993.[2] India suffered from an economic crisis and also a serious balance of payment difficulty associated with declining investment, rising inflation and growing unemployment during this period. In order to overcome the economic crisis, India initiated economic reforms in 1991 with a new economic policy aimed at promoting liberalisation and globalisation of the Indian economy. The economic reforms aimed at loosening the control of the government and encouraged entrepreneurs to participate actively in the process of economic development. This led to the emergence of India's growth story resulting in the rise in GDP at the rate of 8 per cent per annum during 2000s (Ahluwalia, 2011). There was an increase in the rate of migration in urban areas after the Census/NSS year 1991–1993 as revealed by both the census data of 2001 and the NSS year 2000. The increase is also supported by the NSS round for the year 2007–08 both for males and females. However, in subsequent surveys while rate of male migration has declined while for females it continued increasing. It appears that the feminisation of urban migration has become a definite trend in India in recent times.

Migration streams by rural and urban origins are important dimensions of migration trends and patterns. It will be worthwhile examining the trend in migration by various streams and also the categories that reflect upon short- and long-distance migration. As mentioned earlier, migration data by rural and urban status is available and also by administrative locations such as intra-district, inter-district and inter-state. Intra-district and inter-district may be considered as short distance whereas inter-state migration may be treated as long-distance migration. However, in many cases, the destination district in respect of inter-state migration is very close to the state boundaries. As migration data by distance is not available, these proxies are likely to throw some light on the nature of movement which may be characterised as relatively proximate and distant. Table 5.2 shows the composition of migration streams by sex. It may be seen from the table that the rural to urban stream has the largest proportion of male migrants comprising of one-third of the total male migrants observed over almost twenty years' time from 2001 to 2020–21 except in NSSO round 2007–08 showing little higher followed by the rural to rural stream which drastically declined from 31 per cent observed during 2011 census to 18 per cent during PLFS survey in 2020–21. As PLFS 2020–2021 was conducted from July 2020 to June 2021 during COVID-19 times, relatively high level of urban to rural migration has been observed from about 8 per cent in earlier years to about 21 per cent during 2020–2021. Among females, rural to rural migration is the dominant form of migration followed by rural to urban. However, over the years, the percentage of rural to rural female migrants has declined compared to more recent data with that of since 1981. Commensurately some increase is also evident in the rural to urban as well as urban to rural migration. Thus, basically increasing feminisation of migration as observed in Table 5.1 is related to urban transition reflecting increasing proportion of rural females being married in urban areas.

As Table 5.3 shows, the intra-district movement was about 58 per cent in 2020–2021 and, it shows some decline compared to the last census 2011 (about 62 per cent). However, the majority of migrants belong to the intra-district category followed by inter-district and inter-state. Intra-district migration is also relatively higher among females compared to males. It is worthwhile emphasising that internal migration in India is predominantly a 'within state' phenomenon as less than 15 per cent of total migrants moved from one state to another. The states which receive a large number of migrants are Maharashtra, Haryana, Punjab, Gujarat and Karnataka. On the other hand, among outmigration states, Bihar tops the list followed by Uttar Pradesh, Jharkhand and Orissa. The net inter-state migration rate is presented in Fig. 5.1. Maharashtra tops the list among the major states with net in-migration followed by Haryana, while Bihar is the state with the highest net out-migration rate followed by Uttar Pradesh among all duration migrants in 2011.

While inter-state migration comprises less than 15 per cent of all migrants for the combined rural and urban areas and for both sexes, its disaggregation shows as high as one-third male and one-sixth female migrated to urban areas of another state as per more recent data (see Table 5.4).

Further, there was a huge inter-state urban to rural return during the COVID-19 lockdown. It may be observed from Table 5.4 that inter-state male migration to rural areas more than doubled from 16 per cent observed during 2007–08 to about 35 per cent during 2020–2021.

**Table 5.2** Percentage distribution of migrants by streams of migration, India, 2001 to 2020–21

| Streams | Male | | | | | Female | | | | |
|---|---|---|---|---|---|---|---|---|---|---|
| | 2001 census | 1999–2000 NSSO | 2007–2008 NSSO | 2011 census | 2020–2021 PLFS | 2001 census | 1999–2000 NSSO | 2007–2008 NSSO | 2011 Census | 2020–2021 PLFS |
| Rural to Rural | 36.3 | 32.3 | 27.1 | 31.3 | 18.0 | 72.4 | 70.3 | 70.0 | 63.3 | 63.3 |
| Rural to Urban | 34.2 | 34.3 | 39.0 | 30.1 | 33.5 | 13.5 | 14.4 | 14.8 | 15.3 | 15.6 |
| Urban to Rural | 6.2 | 10.7 | 8.9 | 8.6 | 20.8* | 4.1 | 5.2 | 4.9 | 6.0 | 7.8 |
| Urban to Urban | 23.0 | 22.6 | 24.8 | 29.9 | 27.6 | 9.9 | 10.1 | 10.3 | 15.4 | 13.2 |
| Total | 100.0 | 100.0 | 100.0 | 100.0 | 100.0 | 100.0 | 100.0 | 100.0 | 100.0 | 100.0 |

*Source* Census of India 2001 and 2011, D-2 Migration Table; (NSSO, 2010). NSO (2022) Periodic Labour Force Survey (PLFS), 2020–21; *2020–21 refers to the period July 2020–June 2021*. * Urban to rural migration was impacted due to COVID-19 lockdown
*Note* Rural–urban unclassifiable is excluded

**Table 5.3** Percentage distribution of migrants by streams and administrative locations, India, 2001 to 2020–2021

| Census/NSS/PLFS | Intra-district | Inter-district | Inter-state | All |
|---|---|---|---|---|
| Census 2001 | 62.5 | 24.1 | 13.2 | 100 |
| Census 2011 | 61.7 | 26.3 | 12.0 | 100 |
| NSS 1999–2000 | 56.1 | 30.0 | 13.9 | 100 |
| NSS 2007–08 | 53.3 | 32.0 | 14.7 | 100 |
| PLFS 2020–2021 | 58.3 | 29.8 | 11.9 | 100 |

*Source* Census of India 2001 and 2011, Migration Table D-2; NSSO (2001), NSSO (2010), NSO (2022) Periodic Labour Force Survey (PLFS), 2020–21; *2020–21 refers to the period July 2020–June 2021*

## 5.4 Reasons and Characteristics of Urban Migrants

As mentioned previously, migration to urban areas is predominantly rural to urban followed by urban to urban migration. Table 5.5 presents the motives for migration to urban areas. The share of employment-related migration has some increase particularly to take up either employment or better employment for males since 2000, but for females, it is marriage which has been on the rise. Interestingly about 30 per cent of male migrants reported that they migrated to urban areas either in search of employment or better employment and this magnitude has not varied much during the last two decades. However, migration related to movement with parents and earning family members, excluding marriage, has declined but continues to be substantial. It would seem that male and female migration to urban areas is also related as males move for employment and for better employment prospects whereas females follow them as a consequence of marriage or move later as soon as the male migrants settle down. However, the difference between male and female-accompanied migration has almost converged. It indicates that many males also accompany the migrant parents or earning members for various various purposes like searching job, education and for exploring business opportunities (see Table 5.5). It is also significant to mention that education as an independent reason for migration has not increased in more recent years. This may be due to the large expansion of educational institutions in rural areas in recent decades. A very important finding which is not part of the table but emerges from another NSS report (NSSO, 2023, statement 34:48) is that it is the employed persons who have migrated predominantly compared to the unemployed and those out of the labour force.

In India, income data at the household level are generally not available, so studies generally rely on consumer expenditure which includes both food and non-food expenditure. The NSSO provides information on migrants by monthly per capita consumer expenditure (MPCE) of the households. The percentage of migrants in urban areas increases with rising MPCE for both males and females. The rise in the migration rate with MPCE is much steeper for males compared to that for females. The migration rate rises five times for males from the lowest to the highest decile

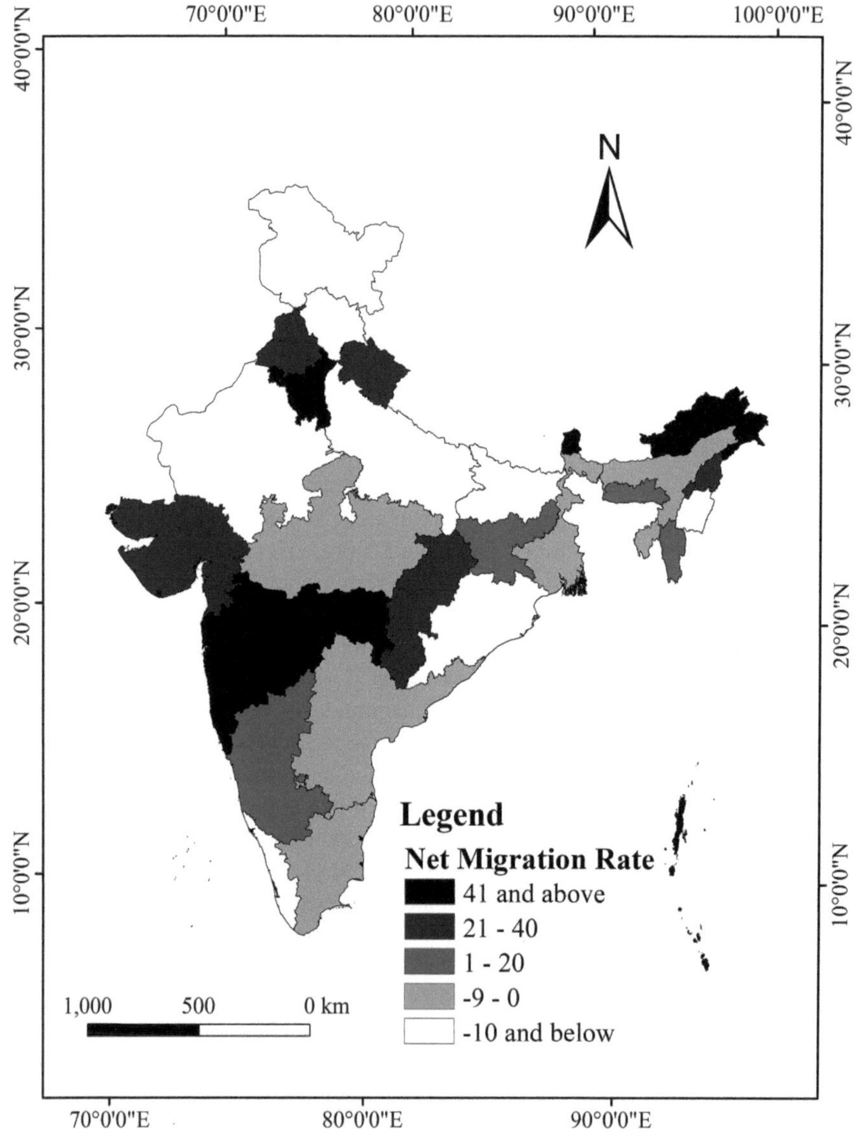


**Fig. 5.1** Net inter-state migration in India, 2011 (all duration) (*Source* Census of India 2011, D-2 Migration Table [www.censusindia.gov.in])

class compared to 1.2 times increase for females. Furthermore, India has a number of communities which are historically disadvantaged due to the caste system. Caste is an endogamous group hierarchically arranged based on birth. As far as migration is concerned, higher castes migrate more than lower castes so far permanent/semi-permanent migration is concerned involving change in the habitual place residence,

**Table 5.4** Migration by administrative locations in rural and urban areas (%), India, 1999–2000 to 2020–2021

| Type by Location | Migrated in Rural Areas | | Migrated in Urban Areas | |
|---|---|---|---|---|
| | Male | Female | Male | Female |
| **NSSO 1999–2000** | | | | |
| Intra-district | 59.7 | 74.6 | 34.5 | 45.1 |
| Inter-district | 25.4 | 20.7 | 40.3 | 38.6 |
| Inter-state | 14.7 | 4.6 | 25.0 | 16.2 |
| Total | 100 | 100 | 100 | 100 |
| **NSSO 2007–08** | | | | |
| Intra-district | 55.6 | 72.1 | 27.9 | 40.6 |
| Inter-district | 27.9 | 23.5 | 38.9 | 40.4 |
| Inter-state | 16.4 | 4.2 | 33.0 | 18.9 |
| Total | 100 | 100 | 100 | 100 |
| **PLFS 2020–2021** | | | | |
| Intra-district | 35.1 | 70.2 | 30.1 | 49.5 |
| Inter-district | 29.8 | 25.7 | 39.2 | 35.4 |
| Inter-state | 34.9* | 4.0 | 30.6 | 15.0 |
| Total | 100 | 100 | 100 | 100 |

*Source* NSSO (2001), (2010). Periodic Labour Force Survey (PLFS), 2020–21. *The year 2020–21 refers to the period July 2020–June 2021.* * Migration to rural areas from urban centres due to COVID-19 lockdown in 2020

while temporary and seasonal migration has been higher among lower castes (Bhagat, 2010; Keshri & Bhagat, 2012, 2013) It is also important to mention that not only do permanent/semi-permanent migrants have a higher level of education compared to the non-migrants but the proportion of educated and technically qualified migrants in urban areas has also risen (Kundu & Saraswati, 2012). There are also numerous studies on India to support the fact that the propensity to migrate increases with rising economic status for permanent and semi-permanent migration in contrast to seasonal, temporary and circulatory migration comprising of people of low economic conditions (Bhagat, 2010, 2014; Keshri & Bhagat, 2012, 2013; Oberai & Singh, 1983; Skeldon, 1986).

The above discussion relates to permanent and semi-permanent migration involving residential change. It does not cover seasonal, temporary and circulatory migration also known as short-term migration without involving residential change. Seasonal, temporary and circulatory migration predominantly occurs among the poor and the socially disadvantaged groups. Temporary migration also largely occurs in the rural to urban migration stream. Further studies show that the annual rate of short-term migration is seven times higher than permanent migration (Keshri & Bhagat, 2013). Studies also point out that that seasonal and temporary migration is a livelihood strategy among rural households (de Haan, 2011; Deshingkar & Farrington,

**Table 5.5** Percentage distribution of migrants by reasons of migration to urban areas, India, 1999–2000 to 2020–2021

| Reasons | Male | | | Female | | |
|---|---|---|---|---|---|---|
| | 1999–2000 | 2007–08 | 2020–2021 | 1999–2000 | 2007–08 | 2020–2021 |
| In search of employment/better employment | 32.6 | 31.6 | 29.9 | 1.5 | 1.2 | 1.5 |
| To take up employment/better employment/transfer of services/proximity to work | 19.3 | 24.1 | 23.6 | 1.5 | 1.4 | 1.7 |
| Marriage | 1.6 | 1.4 | 2.7 | 59.7 | 60.8 | 69.5 |
| Movement with parents and earning family members | 27.0 | 25.2 | 20.8 | 31.0 | 29.4 | 19.4 |
| Studies/Education | 6.2 | 6.8 | 4.5 | 1.3 | 2.2 | 2.4 |
| Loss of job/closure of unit/lack of employment opportunities | – | – | 3.0 | – | – | 0.4 |
| Others* | 13.3 | 13.9 | 15.5 | 6.2 | 5.0 | 5.1 |
| **Total** | **100.0** | **100.0** | **100.0** | **100.0** | **100.0** | **100.0** |

*Source* NSSO (2001), (2012), NSO (2022) Periodic Labour Force Survey (PLFS), 2020–21; *2020–21 refers to the period July 2020–June 2021*
*Note* Others*: This includes business; acquisition of own house or apartment; housing problems; health care; post-retirement; natural disaster and social/political problems

2009; Keshri & Bhagat, 2012). The largest proportion (about 36%) of short-term migration is employed in the construction industry followed by agriculture (20%) and manufacturing (about 16%). There is a dearth of data on the actual magnitude of seasonal, temporary and circular migration and estimates vary from about 13 million, based on the NSSO, to 100 million evaluated by individual researchers (Deshingkar & Akter, 2009; Keshri & Bhagat, 2013). Earlier the Planning Commission suggested that there is a need to undertake state-centric surveys to capture the flow and pattern of migration to the various sectors, particularly the construction sector (Planning Commission, Government of India, 2013: 363). Srivastava opines that seasonal, temporary and circular migration has been increasing in the recent past (2012a).

The National Sample Survey in its 64th Round defined short-term migrants as those people who were away from their village or town for 30 days or more, but for not more than six months in the preceding 365 days, for the purpose of employment or search for employment. This definition also kept on changing and duration varied from two months away for work in the 55th Round to just just 15 days in 70th NSS round (Bhagat & Keshri, 2023). They were enumerated at their place of origin and considered as not having changed their POLR. It is also important to note that commuting as a form of spatial mobility providing an alternative to residential migration is a result of improved transportation between rural and urban areas. Chandrasekhar (2011) estimated that a total of about 32 million individuals live in households where one or more members commute for work from rural to urban areas compared to 15 million commuting from urban to rural areas. Migration to urban areas is, therefore, not only the product of opportunities and constraints in the urban centres but is also influenced by alternative forms of spatial mobility emanating from urban transition.

## 5.5 Opportunities for Urban Ward Migration

Rural to urban migration is not viewed positively in many countries including India. Policies are often aimed at reducing rural to urban migration, although this may not be desirable as there are many reasons to believe that migration, particularly rural to urban, may prove beneficial (Lucas, 1997). The two-sector classical model by Lewis (1954: 139–191) argued that rural to urban migration helps to transfer the surplus labour and boost economic development. In the Indian context, Dubey, Palmer-Jones and Sen (2004) found strong support for the Lewis model. They also observed that as relatively better off sections of the populace and higher castes migrate from rural areas, the gaps that emerge are likely to be filled by the poor and the lower castes with implications for economic improvement and poverty reduction in rural areas. A number of studies show that internal migration can help reduce or prevent households sliding into poverty in both sending and receiving areas (Deshingkar & Grimm, 2005; Harris, 2004; Higgins et al., 2010). The efficiency of labour use and poverty reduction are the two main outcomes associated with the transfer of surplus labour from agriculture to the non-agricultural sector. NSS data also support the increased transfer of the labour force from agriculture to the non-agricultural sector, and migration has emerged as a means to this transformation. Studies also show that migrants (permanent and semi-permanent nature) are better off than non-migrants at their place of destination (Kundu & Sarangi, 2007; Srivastava & Bhattacharya, 2003). Furthermore, the majority of permanent and semi-permanent male migrants were employed in regular jobs in urban areas compared to non-migrant males who were largely self-employed or worked as casual labourers. However, seasonal and temporary migrants fall largely behind when compared to permanent/semi-permanent migrants and non-migrants in urban areas (Srivastava, 2012a). The consequences of migration, therefore, also depend upon the nature and diversity of migration patterns.

In recent years, cities have come to be considered as engines of economic growth as urban areas contribute about 65 per cent of India's GDP (Planning Commission, Government of India, 2008). Migrants help in growth and capital accumulation by providing cheap labour and undertaking many risky and unsafe jobs which natives prefer not to do, although they may face stiff competition or even conflict with the natives. In addition, the seasonal and temporary migrant labour supply is highly flexible in terms of work intensity, payment regimes, working hours and ineffective regulation of labour laws (Srivastava, 2012a). The seasonal, temporary and circular migrants are employed in many leading sectors such as construction, hotel, manufacturing, transport and various service sectors and they contribute at least 10 per cent of India's GDP (Deshingkar & Akter, 2009). Similar economic contribution of migrants is also observed with respect to Global GDP and a plethora of research indicates that migration does not harm either the employment opportunities of natives or their wages (McKinsey & Company, 2016). Generally, the positive impact of migration is ignored particularly its role in improving the economic conditions of many households in the areas of origin and boosting the economy at the destination. However, over time the increasing levels of education and skills act as a driver of migration and ensure upward occupational and economic mobility. Very importantly migration fills gaps in demand for and supply of labour and allocates skilled and unskilled labour in the labour market. Migrant labour is usually cheap, disciplined and willing to do 3D work namely dirty, dangerous and difficult jobs. Further, migration in general and rural to urban migration in particular provides insurance against risks to households in the source areas by sending remittances, increases consumer expenditure and investment in health, education and assets formation. On the other hand, return migration brings knowledge, skills and innovation (these are known as social remittances). In general, migration is an informal process of skill development. It enhances the knowledge and skills of migrants through exposure and interaction with the outside world. New skills are learnt from co-workers and friends at the place of destination. Thus, various studies do corroborate some of these positive aspects of rural to urban migration. However, in popular perception and even among many researchers and policymakers, rural to urban migration is attributed to higher urban poverty, unemployment, crowding, congestion and environmental degradation. It may be worthwhile to emphasise here that these negative aspects of rural to urban migration do influence urban policy and programmes.

Using both secondary and primary data, Jha (2008) concluded that, in India after 1991, outmigration played a decisive role in asset-building in the areas of origin and contributed to poverty reduction through remittances. Chellaraj and Mohapatra (2014) concluded that both internal and international remittances have a poverty-reducing effect and showed that remittances are associated with higher household expenditure on health and education. Similarly, in other parts of the world, urban employment provides an opportunity for rural households to supplement their earnings and to also diversify their sources of income. Remittances from urban employment were found to boost consumption in rural areas and contribute to household savings (IOM, 2005: 15). Increased investment in agriculture has been possible in some countries through migrants' remittances leading to growth in agricultural production. Remittances are also spent on housing, in accessing health care and schooling of the children (Lucas, 1997). The UNDP Human Development Report noted that household remittances are vital in improving the livelihoods of millions of people in developing countries (UNDP, 2009). Available evidence from India shows that about 55 per cent of male outmigrants send remittances. These remittances constitute half of the average household consumer expenditure and are utilised for food, education and health care. About three-quarters of those households which receive remittances spend the amount on food followed by expenditure on health (37.4%), education (31%) and household durables (20.1%). It is therefore evident that remittances are an important means of food security as higher proportions of households with lower socio-economic backgrounds depend on them for such expenditure (Deshingkar & Sandi, 2012). Several studies have pointed out that migration is a family strategy wherein one or more members are employed in urban areas as insurance against distress and crisis; it also improves their creditworthiness (Deshingkar & Sandi, 2012; Lucas, 1997; Stark & Lucas, 1988).

In India, internal migration has a greater role in the economy compared to international migration. Data shows that the internal migrants send remittances twice that of the international migrants (NSSO, 2010). However, it should be borne in mind that migration alone may not improve household conditions as these are influenced by a number of factors. So, while it is reported that remittances from internal migrants increase school attendance, children of seasonal migrants moving with their parents drop out of school (Deshingkar & Sandi, 2012; Smita, 2008). However, it is necessary to emphasise that migration is very intrinsic to the path of human development, but its impact is also place specific (de Haan, 2011).

A publication by UNESCO (2013) points out, migration is an historical process that has shaped human civilisation, culture and development. Along with economic remittances, returning migrants bring a variety of skills, innovation and knowledge to their areas of origin and these are known as social remittances. Migration also enhances the capabilities of the migrants through exposure and interaction with the outside world. Many learn new skills through previous migrants who helped them to migrate and are part of the migrants' social network at the place of destination. In a study in the state of Andhra Pradesh in south India, more than 20 per cent of migrants were able to learn more than two skills (Deshingkar et al., 2009: 84). In fact, migration is an informal process of capability formation and skill development. Cities are places where new migrants accumulate new skills required by modern production sectors. In the long run, migration could play a positive role with the right type of policy and may benefit both the areas of origin and destination (World Bank, 2009).

## 5.6 Challenges of Urban Migration and Policy Issues

Although migration in general and rural to urban migration in particular is conducive to economic and urban transition, there are many challenges confronting migrants in urban areas. A negative attitude towards migration persists as well as a strong hostility towards migrants in spite of the fact that the Indian Constitution guarantees the right to move as a fundamental principle under Article 19. The conflict between natives and migrants has been a core issue since the 1970s. In one of his early writings, Myron Weiner (1978) surveyed the nature of migration and the emergence of the ideology of the *sons of soil* leading to ethnic conflict in different parts of

India. He presented three types of conflict in three distinct regions, namely Assam, Chotanagpur and Hyderabad. In Assam, the reason for conflict was the success of migrants while the natives failed; in Chotanagpur, the cause was unique where tribal members encountered migrants which led to their subjugation and displacement and in Hyderabad, the basis of conflict was the effort to protect the middle-class *mulki* (natives) from the competing migrants. In all three cases, Weiner (1978) argued that the sons of the soil meet the challenge of outsiders by advancing themselves into elitist positions in an effort to oust their non-indigenous competitors. This process has been accelerated rather than diminished by economic development and modernisation. In the 1980s and later, the ideology of nativism further bolstered and spread to other areas, such as Mumbai, Goa and Meghalaya, inciting conflict and violence against migrants (Rajan et al., 2011). The hatred against migrants in the case of Mumbai is more organised and intense as some political parties based on the sons of the soil ideology articulated strongly and voiced threats to the migrants (Hansen, 2001). Discrimination against migrants and their exclusion in India is more subtle and indirect compared to that in China where migration to cities is controlled through the institution of *hukou* (household registration). Persons with rural hukou may enter a city but are excluded from regular urban welfare benefits and social services, such as access to local schools, urban pension plans, public housing and other rights, that are available to people with urban hukou. Rural hukou labour now inundates almost all low-end factory jobs and services that have turned China into a world factory. This segment of migrants provides cheap labour to the cities and is easily exploited. They are not only vulnerable but also excluded from welfare services, urban citizenship and entitlements (Chan, 2012).

In India, all migrants are not equally vulnerable. Permanent and semi-permanent migrants with higher education and skills can withstand the challenges and succeed in becoming members of the urban citizenship. On the other hand, migrants with low education and skills, together with the seasonal, temporary and circular nature of their employment, are more vulnerable and subject to various kinds of exclusions in urban areas. Many of these migrants are excluded from social security programmes such as public distribution of food, access to education and health care and, most importantly, entitlement to housing at the place of destination owing to the absence of identity and residential proof. Social security programmes are place-bound and the implementation of the programmes falls under the purview of the state governments. The inter-state migrants incur more hardships as the portability of social security programmes is a challenging issue; they also face hostility from native residents instilled with the ideology of the sons of the soil (Bhagat 2012b). As per the provisions of Indian constitution, Article 246 states that inter-state migration and inter-state quarantine is under the Union List (81). On the other hand, vagrancy, nomadic and migratory tribes fall under the Concurrent list (15).

Migration is treated more as an issue of governance rather than one of development. Deshingkar and Sandi (2012) argue that there is untapped potential for human development if the positive impacts of migration can be harnessed properly, but this requires accompanying changes in attitude and institutional structures in order to mainstream migration into development planning. There is no dearth of labour laws that deal with workers in general and migrant workers in particular. More recently most of the labour laws were amalgamated into labour codes namely—the Code on Wages, the Industrial Relations Code (IRC), the Code on Social Security (CSS) and the Occupational Safety, Health and Working Conditions Code (OSH) and by Indian Parliament.

The labour laws generally recognise inter-state migrants on the basis of Inter-state Migrant Workmen Regulation Act, 1979 (ISMWRA) which is now synced with the Occupational Safety, Health and Working Condition Code (OSH) 2020 (Rajan & Bhagat, 2023: 141). Another significant Act that exists is known as Building and Other Construction Workers Act 1996 also synced with OSH Code. Under this Act, Building and Other Construction Workers Welfare Board (BOCW) has been constituted by different state governments which collects a cess on the construction of a building and other construction works. The Act makes several provisions for the safety and welfare of construction workers. There is a huge fund lying with the Boards, and a provision for minimum pension for construction workers after the age of 60 is also under consideration (The Economic Times, 2024). A principal flaw in the Act is that it treats construction workers as immobile and does not provide for locational or even inter-sectoral mobility (Srivastava, 2012b). Further, in this act and other acts as well, there is a lack of registration of workers as many of them lack identity and address proofs.[3]

In addition, the provisions of the Inter-State Migrant Workmen (Regulation of Employment and Conditions of Service) Act, 1979, and also the Building and Other Construction Workers (Regulation of Employment and Conditions of Service) Act, 1996, subsumed under OSH Code are now applicable to establishments employing ten or more inter-state migrants unlike five in earlier Acts. As per data available from the Sixth Economic Census, 70% of establishments employ less than six workers. So, the new law even if improved fails to help the inter-state migrant workers, although its scope was expanded to include other than contractor-driven migration (Varma et al., 2020). However, during the COVID-19 pandemic 2020, the central government also announced One Nation, One Ration Card and also rental housing for migrant workers, but many challenges persist in their implementation.

Sustainable Development Goals (SDGs) adopted by the United Nations General Assembly in 2015 recognise the importance of migration more prominently in several of its Goals namely Goal 10: Reduced Inequalities, Goal 8: Decent Work and Economic Growth and Goal 17: Global Partnership and several others unlike Millenium Development Goals. It is an outcome of increasing global research on migration that showed the positive contribution of migration in development, an adaptive strategy to climate change and also its role in fulfilling the aspirations of the people (UNDP, 2009; World Bank, 2009; UNESCO, 2013; Foresight, 2011; World Bank, 2023). However, the policies and programmes integrating migration and migrants at the destination remain weak. Further, as destinations are shaped by the forces of urbanisation, an inclusive urbanisation as a key strategy for migrants' inclusion and integration has been lacking in achieving sustainable development (Rajan & Bhagat, 2021; UNESCO, 2013). Thus, urban policies and programmes are central in the migration policies in general and rural to urban migration policies in particular.

As rural to urban migration provides a large chunk of labour force, the protection of migrant labour rights is crucial for the implementation of labour laws and protection of labour rights (Bhagat, 2019). Migrant labour particularly those belonging to temporary, seasonal and circulatory nature are more vulnerable to exploitation, long working hours, low wages, risky work, and deplorable living and working conditions need to be protected. Labour laws alone will not be adequate, but universal social protection of working class by the state in view of overriding informality, increasing precarity and persistent insecurity in the labour market is most essential (Bharati et al., 2024). The protection of citizenship rights embodied in right to food, right to education, right to health, right to work, right to vote, etc., is the precondition for the protection of labour rights. This will not only promote urban inclusion and creation of urban citizenship, but will also ensure higher efficiency, productivity and well-being in India's aspirational economy. However, the understanding of rural to urban migration mostly looked from the paradigm of 'distress migration' arising out of poverty and the remedy to contain rural to urban migration through rural development programmes such as Mahatma Gandhi National Rural Employment Guarantee Scheme (MGNREGS) is not adequate (Bhagat, 2019; de Haan, 2011; Planning Commission, Government of India, 2013). The implicit assumptions of 'distress migration' negate the very fact that rural to urban migration is also the result of increasing aspiration and ability to migrate as income and educational levels rise in rural areas. The positive values of migration could far outweigh its negative impacts if supported by proper urban policies and programmes.

**Acknowledgements** This chapter draws upon the earlier publication of the first author '*Urban Migration Trends, Challenges and Opportunities in India', Background Paper for World Migration Report 2015: Migrants and Cities,* International Organization for Migration, however, it is extensively revised and updated.

## Notes

1. There is also a category of long–term temporary and circular migrants who live in the destinations particularly in urban areas for more than six months to several years leaving their families behind, visiting them at regular interval and sending remittances. Information on such labour migration is mostly available from specially designed sample surveys.
2. Some demographers are of the opinion that the quality of the 1991 census was affected due to political turmoil during 1991 in several parts of the country (Srinivasan, 1994).
3. The Government of India has launched a biometric identification of the population residing in India under the UID (Unique Identity) Programme also known as AADHAAR. Migrants have the opportunity to enrol and get an AADHAAR card which can be used for identity and residential proof. But as many migrants lack proper documents for identification, a significant step was taken when a Memorandum of Understanding was signed between the National Coalition of Organisations for Security of Migrant Workers, a group of NGOs working with migrant workers, and the Unique Identification Authority of India (UIDAI) for facilitating the inclusion of migrant workers in the UID programme. See: http://uidai.gov.in. The then Governor of the Reserve Bank of India, Mr. Raghuram Rajan, announced that bank accounts would be opened for migrant workers without proof of address (Times of India, 16 August 2014, p. 21).

## References

Ahluwalia, M. (2011). Prospects and Policy Challenges in the Twelfth Plan. *Economic and Political Weekly, 46*(21), 88–105.

Bhagat, R. B. (2010). Internal Migration in India: Are the Underprivileged Migrating More? *Asia Pacific Population Journal, 25*(1), 31–49.

Bhagat, R. B. (2012a). A Turnaround in India's Urbanization. *Asia-Pacific Population Journal, 27*(2), 23–39.

Bhagat, R. B. (2012b). Migrants (Denied) Right to the City. In *National Workshop on Internal Migration and Human Development: Workshop Compendium, Vol. 2, Workshop Papers, UNESCO and UNICEF, New Delhi* (pp. 86–99).

Bhagat, R. B. (2014). '*Urban Migration Trends, Challenges and Opportunities in India', Background Paper for World Migration Report 2015: Migrants and Cities.* International Organization for Migration. Geneva.

Bhagat, R. B. (2019). Urban Migration and Policy Issues. In S. I. Rajan & S. Sumeetha (Eds.), *Handbook of Internal Migration in India* (pp. 429–448). Sage Publication Pvt Ltd.

Bharati, K., Lobo, L., & Kumar, D. (2024). *Urban Daily Labour Markets in Gujarat, Western India.* Routledge, (South Asia Edition).

Bhagat, R. B., & Keshri, K. (2023). Internal Migration and Temporary Labour Circulation in India. In K. James & T. Sekher (Eds.), *India Population Report*. Cambridge University Press. https://doi.org/10.1017/9781009318846.009

Chan, K. W. (2012). Migration and Development in China: Trends, Geography and Current Issues. *Migration and Development, 1*(2), 187–205.

Chandrasekhar, S. (2011). Workers Commuting Between the Rural and Urban: Estimates from NSSO Data. *Economic and Political Weekly, 46*(46), 22–25.

Chellaraj, G., & Mohapatra, S. (2014). *Internal and International Remittances in India: Implications for Household Expenditure and Poverty.* Paper Presented at the *KNOMAD* International Conference on Internal Migration and Urbanization, Dhaka, 30 April-1 May 2014.

Commander, S. (1989). The Mechanics of Demographic and Economic Growth in Uttar Pradesh, 1800–1900. In T. Dyson (Ed.), *India's Historical Demography: Studies in Famine* (pp. 49–72). Disease and Society Curzon Press.

Davis, K. (1951). *The Population of India and Pakistan.* Princeton University Press.

de Haan, A. (2011). *Inclusive Growth? Labour Migration and Poverty in India* (ISS Working Paper No. 513). Institute of Social Studies, The Hague.

Deshingkar, P., & Akter, S. (2009). *Migration and Human Development in India, Human Development* (UNDP Human Development Research Paper, 2009/13). http://hdr.undp.org/en/reports/global/hdr2009/papers/HDRP_2009_13.pdf

Deshingkar, P., & Grimm, S. (2005). Internal Migration and Development: A Global Perspective, *Migration Research Series*, No. 19. International Organization for Migration, Geneva.

Deshingkar, P., & Farrington, J. (2009). A Framework of Understanding Circular Migration. In P. Deshingkar and J. Farrington (Eds.), *Circular Migration and Multilocational Livelihood Strategies in Rural India* Oxford University Press.

Deshingkar, P., Rao, L., Atker, S., & Farrington, J. (2009). The Evolving Pattern of Circular Migration: The Household Surveys in Andhra Pradesh. In P. Deshingkar & J. Farrington (Eds.), *Circular Migration and Multilocational Livelihood Strategies in Rural India* (pp. 58–87). Oxford University Press.

Deshingkar, P., & Sandi, M. (2012). Migration and Human Development in India: New Challenges and Opportunities, National Workshop on Internal Migration and Human Development in India, 6–7 December 2011, Workshop Compendium, Vol. 2, Workshop Papers, UNESCO/ UNICEF, 2012.

Dubey, A., Palmer-Jones, R., & Sen, K. (2004). *Surplus Labour, Social Structure and Rural to Urban Migration: Evidence from Indian Data.* Paper Presented at the Conference on the 50th

Anniversary of the Lewis Model, 67 July, 2004: (www.sed.man.ac.uk/research/events/conferences/documents/Arthur%20Lewis%20Papers/Sen2.pdf accessed on 19th October, 2014).

Foresight. (2011). *Migration and Global Environmental Change: Future Challenges and Opportunities, Final Project Report*, Government Office for Science, London. www.gov.uk/government/uploads/system/uploads/attachment_data/file/287717/11-1116migration-and-global-environmental-change.pdf

Hansen, T. B. (2001). *Wages of Violence: Naming and Identity in Postcolonial Bombay*. Princeton University Press.

Harris, N. (2004). *The Impact of the Reform of International Trade on Urban and Rural Change* (Draft Paper, Development Planning Unit). University College London.

Higgins, K., Birds, K., & Harris, D. (2010). *Policy Responses to the Spatial Dimensions of Poverty* (Overseas Development Institute (ODI) Working Paper, 328/CPRC 168).

Hutton, J. H. (1986). *Census of India 1931: With Complete Survey of Tribal Life and Systems* (Vol. 1). Gian Publishing House.

International Organization for Migration (IOM). (2005). *Migration, Development and Poverty Reduction in Asia*. International Organization for Migration, Geneva.

Jha, V. (2008). *Trickle Down Effects of Inter State Migration in a Period of High Growth in the Indian Economy* (Centre for the Study of Globalisation and Regionalisation, University of Warwick (CSGR) Working Paper 253/08 October, 2008).

Keshri, K., & Bhagat, R. B. (2012). Temporary and Seasonal Migration: Regional Patterns, Characteristics and Associated Factors. *Economic and Political Weekly, 47*(4), 81–88.

Keshri, K., & Bhagat, R. B. (2013). Socio-economic Determinants of Temporary Labour Migration in India. *Asian Population Studies, 9*(2), 175–195.

Kundu, A., & Sarangi, N. (2007). Migration, Employment Status and Poverty: An Analysis across Urban Centres. *Economic and Political Weekly, 42*(4), 299–306.

Kundu, A., & Saraswati, L. (2012). Migration and Exclusionary Urbanization in India. *Economic and Political Weekly, 47*(26 & 27), 219–227.

Lewis, W. A. (1954). Economic Development with Unlimited Supplies of Labour. *The Manchester School, 22* (2), 139–191.

Lucas, R. E. B. (1997). Internal Migration in Developing Countries. In M. R. Rosenzweig & O. Stark (Eds.), *Handbook of Population and Family Economics* (pp. 722–787). Elsevier Science.

McKinsey and Company. (2016). People on the Move: Global Migration's Impact and Opportunities, McKinsey Global Institute (https://www.mckinsey.com/~/media/mckinsey/industries/public%20and%20social%20sector/our%20insights/global%20migrations%20impact%20and%20opportunity/mgi-people-on-the-move-full-report.pdf), Accessed on November 3, 2024.

McNeill, W. (1984). Human Migration in Historical Perspective. *Population and Development Review, 10*(1), 1–18.

National Sample Survey Office (NSSO). (2001). *Migration in India 1999–2000*, Ministry of Statistics and Programme Implementation, Government of India, New Delhi.

National Sample Survey Office (NSSO). (2010). *Migration in India 2007–08*, Ministry of Statistics and Programme Implementation, Government of India, New Delhi.

National Sample Survey Office (NSSO). (2023). Multiple Indicator Survey in India, NSS 78th Round, 2020–2021, Ministry of Statistics and Programme Implementation, Govt. of India, New Delhi (NSS Report No. 589 (78/5.1/1).

National Statistical Office (NSO). (2022). Migration in India, 2020–2021, Periodic Labour Force Survey (PLFS), Ministry of Statistics and Programme Implementation, Govt. of India, New Delhi.

Nehru, J. (1965). *Glimpses of World History*, Asia Publishing House, Bombay (Reprinted in 1965 with 50 Maps by J. F. Horrabin).

Oberai, A. S., & Singh, H. K. M. (1983). *Causes and Consequences of Internal Migration: A Study in the Indian Punjab*. Oxford University Press.

Planning Commission, Government of India. (2008). *Eleventh Five Year Plan (2007–2012), Vol III: Agriculture, Rural Development, Industry, Services and Physical Infrastructure* (pp. 394–422). Oxford University Press.

Planning Commission, Government of India. (2013). *Twelfth Five Year Plan (2012–2017), Vol II: Economic Sectors*. Sage Publication India Pvt Ltd.

Rajan, S. I, Korra, V., & Chyrmang, R. (2011). Politics of Conflict and Migration. In S. Irudaya Rajan (Ed.), *Migration, Identity and Politics* (pp. 95–107). Routledge.

Rajan, S. I., & Bhagat, R. B. (2023). *Researching Internal Migration*. Routledge.

Rajan, S. I., & Bhagat, R. B. (2021, February). Internal Migration in India: Integrating Migration with Development and Urbanization Policies. KNOMAD Policy Brief 12.

Raza, M., & Habeeb, A. (1976). Characteristics of Colonial Urbanization—A Case Study of Satellitic 'Primacy' of Calcutta (1850–1921). In S. Mazoor Alam & V.V. Pokshishevsky (Eds.), *Urbanization in Developing Countries* (pp. 185–218) Osmania University.

Sharma, R. S. (2005). *India's Ancient Past*. Oxford University Press.

Skeldon, R. (1986). On Migration Patterns in India During the 1970s. *Population and Development Review, 12*(4), 759–779.

Smita, S. (2008). Distress Seasonal Migration and its Impact on Children's Education, Research Monograph No. 28, Consortium for Research on Educational Access, Transitions and Equity (CREATE), Pathways to Access. National University of Educational Planning and Administration (NEUPA), New Delhi.

Srinivasan, K. (1994). Sex Ratios: What they Hide and What they Reveal. *Economic and Political Weekly, 29*(51&52), 3233–3234.

Srivastava, R. (2012a). Internal Migration in India: An Overview of its Features, Trends and Policy Challenges. In *National Workshop on Internal Migration and Human Development: Workshop Compendium, Vol. 2, Workshop Papers, UNESCO and UNICEF, New Delhi* (pp. 1–47).

Srivastava, R. (2012b). Internal Migrants and Social Protection in India: The missing link. In *National Workshop on Internal Migration and Human Development: Workshop Compendium, Vol. 2, Workshop Papers, UNESCO and UNICEF, New Delhi* (pp. 166–193).

Srivastava, S., & Bhattacharya, R. (2003). Globalisation, Reforms and Internal Labour Mobility: Analysis of Recent Indian Trends. *Labour and Development, Special Issue on Labour Mobility, 9*(2), 31–55.

Stark, O., & Lucas, R. E. B. (1988). Migration, remittances and the family. *Economic Development and Cultural Change, 36*(3), 465–481.

The Economic Times. (2024). Pension in Works for 8.5 Million Construction Workers. https://economictimes.indiatimes.com/news/economy/policy/pension-in-works-for-8-5-million-construction-workers/articleshow/114703412.cms?from=mdr, Accessed on 7th November, 2024.

United Nations Development Programme (UNDP). (2009). *Human Development Report, Overcoming Barriers: Human Mobility and Development*, United Nations Development Programme, New York. http://hdr.undp.org/sites/default/files/reports/269/hdr_2009_en_complete.pdf

United Nations Educational, Scientific and Cultural Organization (UNESCO). (2013). *Social Inclusion of Internal Migrants in India,* UNESCO, New Delhi. http://unesdoc.unesco.org/images/0022/002237/223702e.pdf

Varma, D., Bharadkar, K., & Mehrortra, R. (2020). https://scroll.in/article/974137/indias-new-labour-codes-fail-migrant-workers-whose-vulnerability-was-highlighted-by-lockdown-crisis, Accessed on 23rd October 2020.

Weiner, M. (1978). *Sons of the Soil: Migration and Ethnic Conflict in India*. Princeton University Press.

World Bank. (2009). *World Development Report: Reshaping Economic Geography*. The World Bank.

World Bank. (2023). *World Development Report 2023: Migrants, Refugees, and Societies*. World Bank. https://doi.org/10.1596/978-1-4648-1941-4

# Chapter 6
# Migration and Population Change in Megacities

## 6.1 Introduction

Although the first human settlements with distinct urban characters existed as early as 3,200 BCE, the change from rural to urban remained abysmally slow for several thousand years. The few towns and cities in different parts of the world contained only a tiny share of the world's population. In fact, until the onset of the Industrial Revolution in Europe, almost the entire world population lived in rural areas (Weeks, 2018: 36). Large-scale changes in production technologies occurred in north-western Europe, mainly England, in the middle of the eighteenth century and, later spread to the rest of Europe and parts of Northern America across the Atlantic Ocean. Australia and New Zealand, which had strong links with the European population, followed suit. The Industrial Revolution brought about a series of transformations in the socio-economic and demographic conditions, including the urbanisation process. The pace of transition from rural to urban, thus, gathered a distinct momentum in the following decades, and the percentage share of the population residing in the rural areas started declining steadily. Improving overall mortality conditions triggered population growth that, in turn, led to heavy rural to urban migration (Hassan, 2020: 51). Immigration also played a vital role in the growing concentration of population in urban centres of the present-day developed world.

The urban population, which constituted only about 3 per cent of the total population towards the end of the eighteenth century, increased to 6.5 per cent in 1850 (Bhattacharya, 2002: 4219; Lowry, 1990: 150). In another hundred years, i.e. by 1950, a little less than 30 per cent of the world's population was already urban, and since then, world urbanisation has grown at an unprecedented pace. Around the same time, urban transition also spread to the less developed parts of the world. By the close of the last century, marginally short of half of the world's population was living in urban areas. The less developed countries began witnessing urban transition only in the latter half of the twentieth century. As nearly seven-tenths of the population came from less developed parts of the world which began experiencing a sudden



R. B. Bhagat and M. I. Hassan, *Urbanisation and Urban Policies in India*, Exploring Urban Change in South Asia, https://doi.org/10.1007/978-981-96-3181-0_6

spurt in overall growth in population, an exceptionally higher rate of urbanisation meant a decided shift in the urban world from developed to less developed countries since the 1950s. The developed world witnessed a drastic fall in its share of the world's urban population from nearly 60 per cent in 1950 to less than one-third by 2000.

The emergence of large cities and growing urban primacy have been the hallmark of the urban transition worldwide since the beginning. The absolute number of large cities above any threshold has increased dramatically, particularly since the middle of the last century, which coincides with the onset of urban transition in the less developed parts of the world. For instance, at the beginning of the nineteenth century, Beijing (then Peking) was the world's only million plus city (Chandler, 1987, quoted in Cohen, 2004: 27). But by the beginning of the twentieth century, the number of million plus cities rose from 16 to 86 by 1950, and over 400 by 2000. The dominance of the developed countries in terms of their contribution to the number of large cities gave way to the developing world after the 1950s. By 1980, there were more 'million plus' cities in the developing world than in the developed one, and the ten megacities in the world were shared equally between the developed and the developing countries (Bhattacharya, 2002: 4220). The latest available estimates indicate that there were 81 'megacities' urban agglomerations (UA) in 2018 (5 million and more) of which as many as 64 are located in less developed countries. While India has only 8 megacities UA with more than 5 million population, China had 19 such cities in 2018 (United Nations, 2019).

Being the land of one of the oldest civilisations, India possesses a long tradition of urban way of life (Hassan, 2020: 57). Even before the urban revolution began in Europe, India was comparatively more urbanised than the Western world. The accounts of foreign travellers in the fifteenth and sixteenth centuries reveal the presence of several large cities and centres of flourishing trade and commerce including port cities in the country (for details please see Crane, 1955). Many of these cities were primarily the 'administrative seats' of numerous states and principalities while many others, which also included port cities, were centres of thriving trade and commerce. However, with the advent of European power for trade and commerce, the regional structure of economy in the country underwent a significant transformation, and a new urban landscape based on newly established port cities emerged that facilitated exploitative trade and commerce of the British (Hassan, 2020: 57). The interactive system that had previously evolved through ages between large number of centres of handicraft, or service or trade and commerce, on the one hand, and their hinterland of primary production, on the other as well as the one between large cities and smaller towns over space were the casualties. Regional inequality sharpened drastically leading to a major shift of population from both the decaying towns and cities of pre-British period and countryside to major port cities.

Most of the major cities of the modern India, for instance, Mumbai, Kolkata and Chennai (previously known as Bombay, Calcutta and Madras, respectively) to name a few, were built by the Europeans. Being located near the coasts, these cities soon grew into big centres of administrative control and economic and commercial activities. The capital of India was shifted from Calcutta (now Kolkata) to Delhi by the British government in 1912. In addition, several other towns were developed in the interior of the country having distinctly closer access to the source of natural resources. New transport networks were developed connecting these centres to the major ports and cities in order to facilitate both export of natural resources and distribution of finished goods imported from Britain. The British also developed several other urban centres that acted as cantonments or hill stations in different parts of the country for their own benefits. Thus, many of the historic cities in the hinterlands of the major port cities lost their position over time, and there occurred a significant shift of population from the hinterland, and also from the pre-existing towns or cities to the respective port cities and other newly developed urban centres.

Transformation of the regional structure of the economy and the emergence of new urban landscape led to growing primacy of select cities. Mumbai and Kolkata achieved the status of 'million plus' cities (population of 'core city' only) in 1911 itself (Venkatrangan, 1989: 310) even at a time when barely one-tenth of the total population in the country was residing in urban areas. Delhi and Chennai acquired the status in 1951 soon after the independence from the British rule in 1947. The number of 'million plus' cities (population of 'core city' only) grew to 11 in 1981; 19 in 1991; 29 in 2001 and 46 in 2011. If urban agglomerations are taken into account the number goes up to 52[1] as per the 2011 census.

Just as elsewhere in the world, in India also more often than not, the built-up area of a statutory town or a city expands beyond the municipal limit over time. Located adjacent to the municipal limit they display strong functional linkages with the main town or city. They possess basic urban features in terms of infrastructure and amenities and are more urban than surrounding rural areas. However, they do not qualify as independent urban units. They are known as 'outgrowths' of a town or city, and together they form part of what is known as 'urban agglomeration' (UA). The Census of India defines urban agglomeration as 'a continuous urban spread constituting a town and its adjoining outgrowths (OGs), or two or more physically contiguous towns together with or without outgrowths of such towns' (RGI, 2018). It is worthwhile to mention here that in the urban agglomeration, the core town or at least one of the two adjoining towns must have statutory status. Here the contiguity in the urban built-up area of the core town (i.e. a statutory town) and its outgrowth as well as of two adjacent towns (one of which must be a statutory town) is the prime criteria in the identification of urban agglomerations. The geographical area of an urban agglomeration, and also its population, is likely to undergo change from census to census depending upon the revision of the municipal boundary of the core town and the emergence of new outgrowth adjacent to its boundary. The Census of India provides data for all urban agglomerations and their constituent units along with the urban centres not included in any urban agglomeration.

Urbanisation is a spatial process apparently concerned with the reorganisation of space manifest in transformation of land use. More often than not, this process of land-use transformation transcends municipal boundaries of towns and cities. Hence, it is essential to take into account the actual spread of urban centres rather than only the part that lies within its municipal limits. The concept of 'urban agglomeration' thus proves quite helpful in capturing reorganisation of space in a holistic manner.

The discussion on growth of select megacities in India in the forthcoming section is based on the population of urban agglomerations rather than the core city alone.

## 6.2 Megacities: Size and Growth

As already noted earlier, there were 52 urban agglomerations with a million plus population each in 2011 of which Greater Mumbai, Delhi and Kolkata are top in the list with a population size of more than 10 million (i.e. 18.39 million, 16.35 million and 14.06 million, respectively) each. These three megacities along with Chennai, Bangalore (new name Bengaluru), Hyderabad, Ahmadabad and Pune are the 8 mega urban agglomerations with more than 5 million population. They are not only very large in terms of size of population but also very densely populated. While average density in Mumbai works out to be 17,297 persons per $km^2$, the corresponding figures for Delhi and Kolkata are 14,349 and 13,311, respectively. These three cities are followed by Chennai and Bangalore with a population of over 8 million each. Although almost identical in terms of population size, Bangalore (11,385 persons per $km^2$) is more densely populated than the Chennai (9280 persons per $km^2$). Next come Hyderabad, Ahmadabad and Pune with a population of 7.68 million, 6.36 million and 5.06 million, respectively. In terms of geographical area, Pune is less than half the size of the other two and hence, the average density of population in the city works out to be as high as 10,059 persons per $km^2$ as against 6,264 and 5,992 persons per $km^2$ in Hyderabad and Ahmadabad, respectively. Further, the density of population goes as high as more than 20,000 persons per sq. km in the core municipal areas of Mumbai and Kolkata. In each of these megacities owning a house in the core is a dream and getting one on rent is very distant. A large proportion of population, for example, as high as about 50 per cent in Mumbai (Municipal Corporation), also lives in slums due to lack of affordable housing. The availability of relatively cheap housing and development of the mass transport helped linking the core to the peripheries which led higher population growth in the peripheries.

Figure 6.1 shows the population size of eight megacities UA with a population above 5 million in 2011[2]. Taken together these eight urban agglomerations accounted for as much as 53 per cent of the population of all million plus cities, and 22.56 per cent of the total urban population in the country in 2011. According to the estimates of the UN, the contribution of these cities to the number of urban dwellers further went up to 24 per cent in 2018 (UN, 2019). These cities, therefore, hold immense significance not only from the economic and political point of view but also from a demographic perspective. They are home to one in every five urban dwellers in the country. Much of economic activities are concentrated in these megacities and they contribute a significant share in India's GDP.

Kolkata is the first megacity in the country. It achieved this status in 1961. Interestingly, the city had grown into an agglomeration of more than 1 million inhabitants as early as the time of the first census enumeration in 1872 (Venkatarangan, 1989: 315). Kolkata was one of the largest cities of the entire British Empire. Developed as

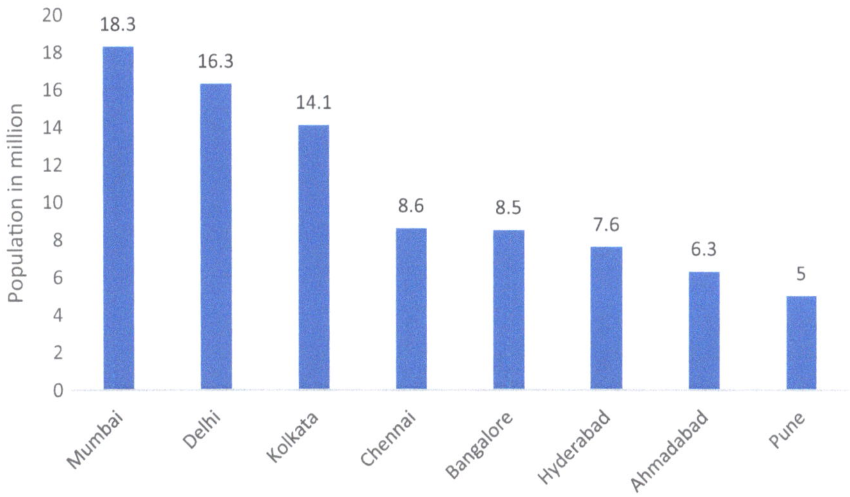


**Fig. 6.1** Population size of eight megacities (UA) (5 million and more), India, 2001–2011 (*Source* Same as in Table 6.1)

a port city for trade and commerce by the East India Company in the late seventeenth and early eighteenth century, Kolkata also served as the capital of the British Raj until 1911. After attaining the status of a megacity in 1961, however, the rate of growth in its population has undergone a continuous decline over time (Table 6.1). The average annual growth rate that remained in the neighbourhood of 2 per cent till the end of the 1990s, declined to less than 1 per cent thereafter. Nevertheless, the dominance of Kolkata on the urban landscape of eastern India continues and it remains the only megacity in the region.

Mumbai became a megacity a decade later in 1971. However, with much faster growth in its population, it soon became the top-ranking city in the country in 1981 pushing Kolkata to second position. Initially a territory under Portuguese control till the middle of the seventeenth century, Mumbai was later developed into a major port city by the British on the western coast. Mumbai is the financial and commercial capital of India with its significant contributions to the country's industrial output, marine trade, capital transactions and national income. On average, the city recorded an annual rate of growth of over 3 per cent till 1981. Thereafter, there has been a gradual decline in the pace of growth in its population. During 2001–11, the average annual rate of growth of Mumbai UA was 1.1 per cent and the same rate continued during 2015–20 as well (Tables 6.1 and 6.2).

Delhi was made the capital city by the British in 1911, and thereafter, it has grown by leaps and bounds. Over one hundred years, i.e. since the time it became the capital city of the British Crown up to 2011, the population of Delhi has expanded by as many as 69 times. It qualified as a million plus city in 1951 and recorded a four-fold increase in its population in just three decades to attain the status of a megacity in 1981. Like Mumbai and Kolkata, Delhi has also undergone a decline in the pace

**Table 6.1** Trends of population growth of megacities (UA) with a population of 5 million and above, India, 1961–2011

| Cities (UA) | Population (million) | | | | | | Average annual rate of change (per cent)[a] | | | | |
|---|---|---|---|---|---|---|---|---|---|---|---|
| | 1961 | 1971 | 1981 | 1991 | 2001 | 2011 | 1961–71 | 1971–81 | 1981–91 | 1991–01 | 2001–11 |
| Kolkata | 6.01 | 7.45 | 9.23 | 11.11 | 13.25 | 14.06 | 2.14 | 2.15 | 1.85 | 1.76 | 0.59 |
| Mumbai | 4.51 | 6.60 | 9.42 | 12.60 | 16.42 | 18.40 | 3.79 | 3.56 | 2.90 | 2.66 | 1.12 |
| Delhi | 2.36 | 3.65 | 5.76 | 8.47 | 12.90 | 16.35 | 4.35 | 4.57 | 3.85 | 4.20 | 2.37 |
| Chennai | 1.94 | 3.16 | 4.23 | 5.42 | 6.69 | 8.65 | 4.84 | 3.03 | 2.37 | 2.10 | 2.57 |
| Bangalore | 1.21 | 1.66 | 2.92 | 4.14 | 5.70 | 8.52 | 3.21 | 5.62 | 3.47 | 3.20 | 4.02 |
| Hyderabad | 1.25 | 1.81 | 2.61 | 4.34 | 5.75 | 7.68 | 3.69 | 3.62 | 5.10 | 2.81 | 2.87 |
| Ahmadabad | 1.27 | 1.87 | 2.73 | 3.63 | 4.91 | 6.36 | 3.91 | 3.79 | 2.83 | 3.02 | 2.57 |
| Pune | 0.79 | 1.14 | 1.72 | 2.49 | 3.77 | 5.06 | 3.61 | 4.16 | 3.70 | 4.12 | 2.94 |
| All[b] | 19.34 | 27.34 | 38.62 | 52.2 | 69.39 | 85.08 | 3.46 | 3.45 | 3.01 | 2.84 | 2.03 |

*Note* [a] Refers to cities (UA) with 5 million or more inhabitants
[b] Exponential growth rate
*Source* Census of India, 2011, *Table A 4 Towns and urban agglomerations classified by population size class in 2011 with variation between 1901 and 2011*

**Table 6.2** Trends of population growth of megacities (UA) with population of 5 million and above, India, 1990–2030[b]

| Cities (UA) | Population (million) | | | World Rank | | | Avg. annual rate of change (per cent) |
|---|---|---|---|---|---|---|---|
| | 1990 | 2018 | 2030 | 1990 | 2018 | 2030 | 2015–20 |
| Delhi | 9.38 | 28.51 | 38.94 | 13 | 2 | 1 | 3.16 |
| Mumbai | 12.36 | 19.98 | 24.57 | 6 | 7 | 6 | 1.10 |
| Kolkata | 10.97 | 14.68 | 17.58 | 8 | 16 | 16 | 0.58 |
| Bangalore | 4.04 | 11.44 | 16.23 | 42 | 27 | 21 | 3.90 |
| Chennai | 5.33 | 10.46 | 13.81 | 30 | 31 | 26 | 2.51 |
| Hyderabad | 4.19 | 9.48 | 12.71 | 39 | 36 | 29 | 2.80 |
| Ahmadabad | 3.55 | 7.68 | 10.15 | 54 | 45 | 43 | 2.51 |
| Surat[a] | 1.47 | 6.56 | 9.71 | 171 | 56 | 46 | 4.73 |
| Pune | 2.43 | 6.76 | 8.44 | 87 | 59 | 55 | 2.86 |

*Notes* [a] With a population of 4.59 million in 2011 Surat did not figure among in the list of the megacities. It was the 9th largest urban agglomeration. The population of the city grew at an annual rate of growth of 6.3 per cent during the preceding decade

[b] Refers to cities with 5 million or more inhabitants in 2018

*Source* United Nations, Department of Economic and Social Affairs, 2019, *World Urbanization Prospects, 2018: Highlights*, Table II pages 29–30

of its growth after the 1990s. Notwithstanding a general deceleration during 2001–11, Delhi replaced Kolkata as the second-largest megacity in India in 2011. In all likelihood, with an average annual rate of growth that is twice that of Mumbai, Delhi is expected to have surpassed the former long back as the largest megacity of India. Moreover, given the present rate of growth, both Mumbai and Delhi are expected to be agglomerations of more than 20 million inhabitants by now, the former was only marginally short of 20 million mark in 2018 (see Table 6.2). They, thus, represent what is known as "meta" or "hyper" cities of the world. Kolkata UA which is growing at an annual rate of less than 1 per cent will have to wait for another decade or more to reach this level. The world ranking of Kolkata UA, according to the estimates of the United Nations (Table 6.2), has slipped down from 8th in 1990 to 16th in 2018.

About its demographics, Delhi is placed quite in a unique situation. Adjacent to its boundaries are located some of the fastest-growing satellite towns/cities like Gurgaon, Faridabad and Bahadurgarh from Haryana, and Greater NOIDA and Ghaziabad from Uttar Pradesh. They provide contiguity in the spread of urban land use. However, these satellite towns/cities are not included in the Delhi urban agglomeration by the Census of India as they fall beyond the state boundary. The United Nations, however, treats this whole expanse as a single urbanised area as a result of which Delhi with an agglomeration of 29 million inhabitants in 2018 figures as the second-largest megacity of the world after Tokyo with 37 million (Table 6.2). With the present rate of growth in its population Delhi is expected to become the largest city of the world anytime by the close of this decade (UN, 2019).

The next to be added to the list of megacities was Chennai in 1991. Developed by the British as a trading post on the Coromandel Coast of the Bay of Bengal, Chennai has emerged as a premier centre of industries and commerce after independence. Software development and electronics manufacturing have formed a significant portion of the economy of the city. Other prominent industrial activities relate to the manufacture of automobiles, rubber, fertiliser, leather, iron ore, cotton textiles, etc. Unlike the other three megacities, Chennai is reported to have witnessed an acceleration in the pace of its growth in 2001–11 as compared to the previous decade. The city grew at an average annual rate of nearly 2.5 per cent per annum during 2001–11 compared to 2.1 per cent during 1991–2001. It continued growing at the same rate during 2015–20 as well as estimated by the United Nations (Table 6.2).

Bangalore and Hyderabad joined the group of megacities in 2001 with almost identical size of population. With the presence of several high-technology-based heavy industries, and top-tier engineering and research institutions, Bangalore is one of the fastest-growing megacities in the world. Of late, it has become the leading exporter of software and is known as the "Silicon Valley" of India. Likewise, along with the presence of traditional industries, Hyderabad has also emerged as an important hub of information technology and IT-enabled services.

However, of the two, Bangalore had an edge over Hyderabad. It grew at a faster pace and gave close competition to Chennai in terms of size of population by 2011. Finally, by the year 2018, Bangalore overtook Chennai and became the fourth-largest megacity in the country (Table 6.2). In the world ranking of cities in terms of size of the population, Bangalore occupied 42nd position in 1990 which changed to 27th in 2018. By 2030, it is projected to be the 21st largest city in the world. Ahmadabad and Pune attained the status of megacity only in 2011. During 2015–20, the average annual growth rates of the two cities were 2.5 per cent and 2.9 per cent, respectively (Table 6.2). At the time of the 2011 census, Surat, a port city in the state of Gujarat, was very close to 5 million inhabitants. It is worth noting that the population of Surat had recorded a growth rate of about 5 per cent per annum during 2001–11. Although the pace of growth declined thereafter, Surat saw an increase in its population at an annual rate of 4.7 per cent during 2015–20 and, by 2018 it acquired the status of a megacity taking the total number to nine (UN, 2019).

It is worth mentioning that barring only Chennai, Bangalore and Hyderabad, all the other megacities have witnessed a slowdown in the pace of growth during 2001–11 over the preceding decade. Moreover, in the case of Hyderabad, there is only a marginal increase in the rate of growth.

Higher growth rates in some of megacity UAs such as Bangalore, Hyderabad and Chennai during 2001–11 compared to 1991–2001 cannot be attributed to migration alone but there has been a significant geographical expansion of the urban agglomeration due to the emergence of new towns or outgrowths and reorganisation of municipal boundaries due to merger of existing urban centres as well (Fig. 6.2). It is interesting to notice a slowing down of the growth rate in Ahmadabad between 2001–11 and 1991–2001 in spite of huge areal expansion of Ahmadabad UA and a moderate increase in area of Delhi UA. We also find decline in area more pronounced

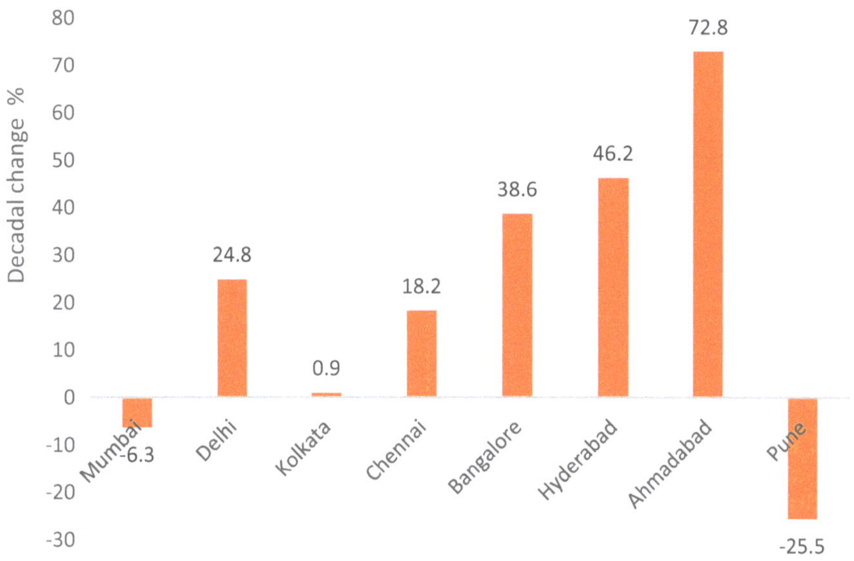


**Fig. 6.2** Decadal area change in megacities (UA), India, 2001–2011 (in per cent) (*Source* Same as in Table 6.1)

in Pune UA which may be responsible for some decline in growth rate during 2001–11 compared to 1991–2001 (See Fig. 6.2). It is important to emphasise that the dynamics of expansion or contraction of an UA depends upon the political process of drawing and redrawing of municipal boundaries of the existing urban units or the emergence of new towns in the UA space in the respective states.

At the UA level, megacities UA like Delhi, Kolkata, Hyderabad, Ahmadabad and Mumbai have shown a decline in their growth rates compared to increase in Bangalore, Hyderabad and Chennai during 2001–2011 compared to 1991–2001 (see Table 6.1). The most perceptible decline in the pace of growth is seen in case of Kolkata followed by Mumbai. This is despite the fact that the geographic area of Kolkata agglomeration has increased albeit marginally between 2001 and 2011. Remarkably, the population in the core city of Kolkata (i.e. Kolkata M. Corp.) has even begun shrinking. The shrinking in population has also been happening in Mumbai Island (Mumbai district) and New Delhi (M.Corp) as well. The area of Mumbai agglomeration did witness shrink during the same period. Next comes Delhi in terms of slowdown in the pace of growth despite an expansion in the geographic area by almost 25 per cent between 2001 and 2011. While explaining growth rate of cities and UAs researchers need to keep this aspect in mind. However it may be noted that adjustment of population data in respect with boundary changes in cities and UAs is cumbersome. Researchers should not jump to the conclusion that cities have only grown due to migration looking at the population growth rates only.

Most importantly the core areas (municipal Corporation areas) of the megacity UAs have been showing a lower growth rate compared to the peripheral areas.

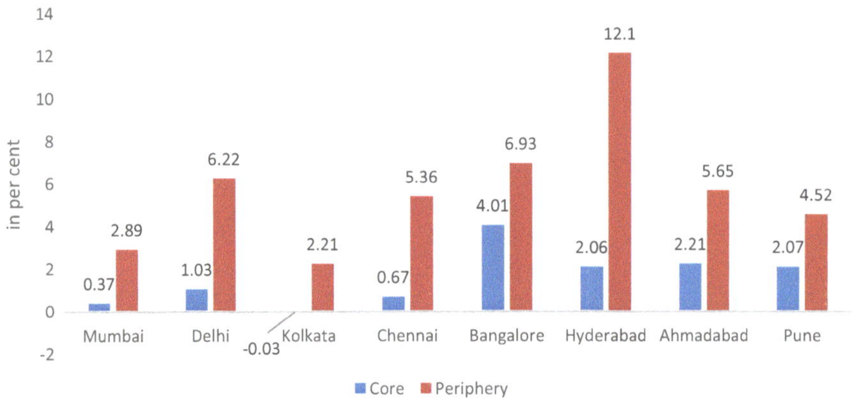


**Fig. 6.3** Average annual population growth rate (%) in Mega UA by core and periphery, 2001–2011 (*Note* Core is defined as the main Municipal Corporation in the respective UA. In the case of Kolkata, it is Kolkata and Haora (Howrah) (M.Corp) together and Delhi Core consists of Delhi (M.Corp) and New Delhi (M. Corp.) *Source* Same as in Table 6.1)

For example, in Mumbai UA, cities like Navi Mumbai, Thane, Kalyan and Mira Bhayander show higher growth rates compared to Municipal Corporation of Greater Mumbai. Similar is true for other mega UAs as well. Most of the peripheries have grown much faster than the core in all megacities UA. In the case of Kolkata, the core has even declined and Mumbai and Chennai show very low population growth in their core areas. Figure 6.3 confirms the findings that the India's urbanisation is not only concentrated but extensive and differential as well as observed in Chapter 4.

## 6.3 Age-Sex Composition

Age and sex are important biological characteristics of population that have significant bearings on not only the other demographic attributes like fertility and mortality but also the social, economic and political aspects of the society. Age structure of a population can be analysed in a number of ways, the simplest of which is the percentage distribution of population in three broad age groups representing children (under age 15 years), adults (between ages 15 and 59 or 64 years) and aged dependents (60 or 65 years and above). Since adults constitute what is generally known as 'economically active' population, a ratio between dependents (i.e. children and aged dependents put together) and adults serves a very useful purpose in understanding the economic conditions of people living in a region. This ratio is known as 'dependency ratio'. Likewise, sex composition of a population is an important indicator of the nature of gender relations in a society and is analysed in terms of a ratio between the numerical strengths of females and males (or vice-versa) in the population. This is known as sex ratio. In the Indian context, sex ratio is defined as number of females per 1000 males in the population. The overall sex ratio of population in a region is the outcome of past trends in sex ratio at birth and death, and among net migrants. Migrants constitute a significant portion of population in large towns and cities in India. In Indian context, migration is highly sex selective in nature, and hence, overall sex ratio is largely determined by spatial mobility of population. Overall sex ratio is rendered meaningless with respect to regional variations in the nature of gender relations. Demographers, therefore, suggest such ratio for some of the early age brackets where sex ratio at birth and sex differential in mortality remain the main determinants and the contribution of migration is minimal. In India, this ratio is generally worked out for the age group 0–6 years for which data are available down up to village level. This is popularly known as child sex ratio. The discussion in the forthcoming paragraphs is based on these measures.

With some notable exceptions, the age structure of population appears to be almost identical across these eight megacities. Children constitute more than a quarter of the total population in Hyderabad, Delhi and Ahmadabad. As a result, these three cities occupy the top positions in the same order in terms of dependency ratio as well. In the rest of the cities, the proportion of children remains below 25 per cent with Kolkata at the bottom with around 19 per cent only. Adults in the working age group from 67 per cent to 69 per cent of the population. Likewise, aged dependents form less than one-tenths of the population except in Kolkata where nearly 12 per cent of the population is above 60 years of age. This variation in age structure is amply reflected in the dependency ratio of these cities. Youth in the age group 15 to 29 years form, on average, more than 40 per cent of the adult population, the highest being in the case of Hyderabad, Bengaluru and Delhi. In Kolkata, however, youth form only 38 per cent of the population in the working age group. Outmigration of youth can be a possible reason for this which led to a distinctly higher proportion of elderly persons. Chennai is only marginally ahead of Kolkata in terms of the proportion of youth among adults. Incidentally, Chennai also ranks very high in terms of the share of elderly in the population.

Large cities attract substantial migrants from outside, and migration to urban areas is highly sex selective. An adverse all-age sex ratio in these cities is a matter of rule rather than an exception. On average, there are only 904 females per thousand males in these megacities as compared to the national average of 929 for the urban areas[3]. In the case of Greater Mumbai, there are only 853 females per thousand males. In addition to Greater Mumbai, in Delhi and Ahmadabad also all-age sex ratio is well below 900. Interestingly, however, in Chennai there are as many as 989 females per thousand males in the population. Remarkably, barring only Chennai, Hyderabad and Pune, all other megacities are marked with a lower all-age sex ratio than the all-India average. The child sex ratio, on the other hand, at the aggregate level for these megacities is almost identical to all-India average for the urban areas. An extremely adverse sex ratio among children is seen in the case of Ahmadabad and Delhi both located in the north-western parts of the country with a significant deficit of female children in the population. On the other hand, cities located in the southern and eastern parts of the country report much better child sex ratio, the highest being in the case of Chennai followed by Bengaluru and Kolkata (Table 6.3).

**Table 6.3** Age structure and sex composition of population of megacities, India, 2011

| Cities | Per cent population in age groups | | | Youth[a] as % to adults[b] | Dependency Ratio | Sex Ratio | |
|---|---|---|---|---|---|---|---|
| | 0–14 | 15–59 | 60 + | | | All ages | 0–6 years |
| Kolkata (M Corp.) | 18.96 | 69.21 | 11.83 | 38.02 | 44.49 | 908 | 933 |
| Mumbai (M Corp.) | 22.15 | 69.25 | 8.60 | 43.63 | 44.41 | 853 | 913 |
| Delhi[c] | 25.37 | 66.96 | 7.67 | 44.75 | 49.35 | 874 | 871 |
| Chennai (M Corp.) | 21.52 | 68.62 | 9.85 | 39.98 | 45.73 | 989 | 950 |
| Bangalore (M. Corp. + OG) | 22.81 | 69.61 | 7.58 | 45.33 | 43.66 | 922 | 942 |
| Hyderabad (M Corp. + OG) | 26.80 | 66.49 | 6.71 | 46.37 | 50.39 | 955 | 921 |
| Ahmadabad (M Corp. + OG) | 25.05 | 66.97 | 7.98 | 43.37 | 49.32 | 898 | 848 |
| Pune (M Corp.) | 23.13 | 67.68 | 9.19 | 43.64 | 47.75 | 948 | 908 |
| All | 23.49 | 68.15 | 8.36 | 43.66 | 46.74 | 904 | 906 |

*Notes*
[a] Refers to age group 15–29 years
[b] Refers to age group 15–59 years. Dependency ratio = (0–14 + 60 plus)/15–59 * 100
[c] Includes Delhi Municipal Corporation, New Delhi Municipal Corporation and Delhi Cantonment
[d] M. Corp—Municipal Corporation; OG—Outgrowth
*Source* Compiled from Census of India 2011, *Social and Cultural Tables* Table-C08-CITY, Educational level by age and sex for population age 7 and above (cities)—2011, https://censusindia.gov.in/census.website/data/census-tables, accessed on 18 November 2024

## 6.4 Literacy and Education

Literacy and education play a crucial role in the social and economic development of a country. Urban areas are generally ahead of their rural counterparts in terms of literacy rate and levels of educational attainment. Megacities are hubs of economic and commercial activities. They act as an important link with the world economic system through business, trade and transport. They contribute a disproportionately larger share of the national GDP. They offer distinct prospects and challenges for the job seekers and, therefore, levels of literacy and educational attainment in these megacities are markedly different from the rest of the urban centres. As against an average literacy rate of 84 per cent for urban India, in the eight megacities as a whole literacy rate stands at 88 per cent. Chennai is the most literate megacity closely followed by Greater Mumbai and Pune. The lowest literacy rate among these megacities is found in Hyderabad. Interestingly, the literacy rate in Hyderabad is even lower than the average of urban India. A comparatively lower literacy rate can be attributed to its religious composition of population. Muslims who report the lowest literacy rates among all religious groups account for nearly one-third of the population in the city (Kumar, 2014).

The relative position of the megacities with regard to educational attainment, however, is different. Delhi and Pune emerge as top-ranking cities with regard to the proportion of 'graduates and above' in the population aged 20 and above (see Table 6.4). These two cities have long been the hubs of educational institutions. Bengaluru that has made rapid strides after independence and has emerged as an important centre in the field of information technology comes next to them. Chennai falls behind Bengaluru only by a small margin. Ahmadabad and Greater Mumbai occupy the lowest rungs among the megacities with respect to the share of 'graduates and above' in the population. It may be noted that Greater Mumbai is known for a very large share of population residing in slums a large chunk of which depend on the informal sector for survival. This could be a plausible reason for a relatively small share of graduates in the population. Likewise, the economy of Ahmadabad is largely dependent on business and trade and it is quite likely that youth prefer to join their traditional occupations after schooling rather than continuing their higher education. This aspect shall become clear when workforce characteristics in these megacities are discussed later in the chapter.

**Table 6.4** Literacy rates and education among megacities, India, 2011

| Cities | Literacy Rate[a] | % Graduates and above[b] |
|---|---|---|
| Kolkata (M Corp.) | 86.31 | 25.06 |
| Mumbai (M Corp.) | 89.73 | 21.01 |
| Delhi[c] | 87.67 | 31.03 |
| Chennai (M Corp.) | 90.18 | 26.83 |
| Bangalore (M. Corp. + OG) | 88.68 | 27.06 |
| Hyderabad (M Corp. + OG) | 82.93 | 26.09 |
| Ahmadabad (M Corp. + OG) | 88.26 | 20.82 |
| Pune (M Corp.) | 89.56 | 30.62 |
| All | 87.96 | 25.06 |

*Notes* [a] Literates as percentage of population aged 7 and above
[b] Refers to the age group 20 years and above
[c] Includes Delhi Municipal Corporation, New Delhi Municipal Corporation and Delhi Cantonment
*Source* Same as Table 6.3

## 6.5 Migration

Just as elsewhere in the world, the population residing in urban areas in India is more mobile than its counterpart in rural areas. As per the 2011 census, barely one-third of the population in rural areas was migrants. As compared to this over 46 per cent of the population in the urban centres was classified as migrants. Migration has played a significant role in the process of urbanisation in the country. Migration to urban centres generally occurs for a variety of reasons. Besides employment, better job opportunities, higher wage levels, improved conditions of life, better educational infrastructure and health services are some of the important reasons for migration. In addition, job transfer, marriages and a host of personal factors including the movement of the family from one location to another also affect migration to urban areas. Urban centres vary a great deal in terms of these factors underlying the process of migration. Large cities are generally more endowed with these opportunities that attract migrants, and hence, a marked correspondence between the size class category of urban centres and the rate of mobility is an inherent feature of population mobility. As Kundu (2007) suggests migration in the country has become more concentrated in specific areas because new job opportunities are available in selective sectors and urban centres, at the same time a gradual shift in the migration from megacities to other smaller cities is also taking place in developing countries including India (Marchang, 2022: 57).

These eight megacities in the country on the whole were the home of 43.17 million migrants at the time of the 2011 census. This accounted for more than half of their population at the aggregate level. The agglomerations of Kolkata, Greater Mumbai and Delhi are the three oldest megacities of the country. Taken together they alone were the destinations for more than 53 per cent of all migrants in the eight megacities. If only inter-censal migrants[4] are taken into account, the corresponding figure comes down to 50 per cent which confirms the shift in migration away from these cities after 2001. However, this shift is more marked in the case of Kolkata than Greater Mumbai and Delhi. Bengaluru and Pune, on the other hand, seem to be gaining part of this shift as their shares in migrants and inter-censal migrants indicate. Hyderabad and Pune which attained the status of a megacity only in 2001 and 2011, respectively, report nearly 65 per cent of its population as migrants.

Inter-state migrants account for more than one-third of all migrants in these megacities. Delhi urban agglomeration is the most favoured destination for migrants from across the state boundaries followed by Greater Mumbai and Bengaluru. Nearly nine-tenths of all migrants in Delhi urban agglomeration have come from other states/ union territories. It is argued that Delhi being the national capital provides more job and education opportunities and hence this large share of inter-state migrants (Marchang, 2022). However, this is a very simplistic explanation that glosses over the unique position of Delhi among all other megacities. The administrative boundary of Delhi's urban agglomeration is coterminous with the boundary of NCT of Delhi. Hence, any in-migration from across the administrative boundary of Delhi urban agglomeration is technically classified as 'inter-state' migration. This is not the case with other megacities where a good part of such in-migration falls under 'intra-state' category. The major sending states for in-migrants to Delhi are Uttar Pradesh and Bihar. Uttar Pradesh alone contributes over four-tenths of migrants to Greater Mumbai also. The other major sending states for Greater Mumbai are Gujarat and Karnataka. For Kolkata urban agglomeration also Bihar and Uttar Pradesh are major sending states contributing nearly seven-tenths of migrants. Bengaluru on the other hand draws migrants mainly from southern states for instance Tamil Nadu, Andhra Pradesh and Kerala. More than two-thirds of all migrants in Bengaluru come from these states. Chennai too draws a major part of migrants from the southern states mainly. Hyderabad and Ahmadabad also draw migrants from states adjacent to their respective state boundaries.

Interestingly, inter-state migrants constitute a larger proportion of all migrants in the city proper, i.e. core city than in the periphery, i.e. the constituent parts of the agglomeration in the case of Kolkata, Greater Mumbai, Chennai and Ahmadabad while in others the shares are almost identical. In other words, while the core remains the main attraction for migrants from other states, migrants from within the state boundary chose to settle mainly in the periphery. The location quotients, a measure of population concentration, presented in Table 6.5 further show that when all migrants (intra state + inter-state) are taken into consideration migrants are more concentrated in peripheries of most megacity UA. This is consistent with the higher population growth rate in the peripheries of most of the megacity UAs compared to their cores (Table 6.6).

**Table 6.5** Percentage of migrants and share of inter-state migrants in the megacities UA of India, 2011

| Cities (UA) | Migrants[1] as % to total population | | Inter-state migrants as % to all migrants | | Location Quotient[2] |
|---|---|---|---|---|---|
| | UA | Core City[a] | UA | Core City[a] | |
| Kolkata | 40.98 | 18.62 | 18.24 | 47.05 | 2.20 |
| Mumbai | 54.98 | 44.24 | 45.92 | 62.65 | 1.24 |
| Delhi | 43.07 | 40.60 | 87.82 | 87.60 | 1.06 |
| Chennai | 51.30 | 29.50 | 11.77 | 25.50 | 1.74 |
| Bangalore | 52.22 | 52.46 | 35.15 | 35.14 | 1.00 |
| Hyderabad | 64.98 | 44.25 | 7.08 | 7.02 | 1.47 |
| Ahmadabad | 48.73 | 45.97 | 24.12 | 27.14 | 1.06 |
| Pune | 64.76 | 58.00 | 22.39 | 22.78 | 1.12 |
| All | 50.74 | 42.43 | 36.60 | 46.02 | 1.20 |

[a] Core refers to main Municipal Corporation of UA. Delhi core includes both Delhi (M.Corp) and New Delhi (M.Corp.)

1. Migrants refer to all duration migrants
2. A ratio between two ratios—the first ratio is between the absolute number of migrants in the urban agglomeration and in the core city, and the second is between the absolute size of population in the urban agglomeration and in the core city

*Source* Based on Census of India 2011, *Table D-3 (UA): Migrants by Place of Last Residence, Duration of Residence and Reasons of Migration,* https://censusindia.gov.in/census.website/data/census-tables#, accessed on 18 November 2024

**Table 6.6** Major source regions for inter-state migrants in megacities, India, 2011

| Cities (UA) | Top five sending states |
|---|---|
| Kolkata | Bihar, Uttar Pradesh, Jharkhand, Odisha and Rajasthan |
| Mumbai | Uttar Pradesh, Gujarat, Karnataka, Rajasthan and Bihar |
| Delhi | Uttar Pradesh, Bihar, Haryana, Rajasthan and Uttarakhand |
| Chennai | Andhra Pradesh, Kerala, Karnataka, Rajasthan and Maharashtra |
| Bangalore | Tamil Nadu, Andhra Pradesh, Kerala, Rajasthan and Maharashtra |
| Hyderabad | Karnataka, Maharashtra, Tamil Nadu, Rajasthan and Uttar Pradesh |
| Ahmadabad | Rajasthan, Uttar Pradesh, Maharashtra, Madhya Pradesh and Bihar |
| Pune | Karnataka, Uttar Pradesh, Rajasthan, Andhra Pradesh and Bihar |

*Source* same as in Table 6.5

## Notes

1. Chandigarh that was included in the list of million plus cities according to the provisional figures of the 2011 census was later excluded in the Final Totals.
2. According to the estimates of the United Nations (UN, 2019), with a population of 6.56 million Surat is the 9th megacity in the country.

3. The child sex ratio for the urban population in the country works out to be 906 females per 1000 males as per the 2011 census.
4. Who has moved to these megacities after 2001 census.

## References


Bhattacharya, P. C. (2002). Urbanisation in Developing Countries. *Economic and Political Weekly, 37*(41), 4219–4228.

Chandler, T. (1987). *Four Thousand Years of Urban Growth: A Historical Census.* St. David's University Press.

Cohen, B. (2004). Urban Growth in Developing Countries: A Review of Current Trends and a Caution Regarding Existing Forecasts. *World Development, 32*(1), 23–51.

Hassan, M. I. (2020). *Population Geography: A Systematic Exposition.* Routledge.

Kumar, J. (2014). Progress of Urban Literacy in India: Focus on Metropolises. *Indian Journal of Geography and Environment, 13*, 70–77.

Kundu, A. (2007). Mobility of Population. In K. Basu (Ed.), *Oxford Companion to Economics in India* (pp. 443–48). Oxford University Press.

Lowry, Ira S. (1990). World Urbanization in Perspective. *Population and Development Review*, Vol. 16, Supplement: Resources, Environment and Population: Present Knowledge, Future Options, pp. 148–176.

Marchang, R. (2022). Emerging Pattern and Trend of Migration in Megacities. *Economic and Political Weekly, 57*(15), 56–61.

Registrar General of India. (2018). *Formation of Urban Agglomerations for the 2021 Census, Census of India—Circular No. 3.* https://www.censusindia.gov.in/census.website/data/CCIRC. Accessed on October 09, 2023.

United Nations, Department of Economic and Social Affairs, Population Division. (2019). *World Urbanization Prospects 2018: Highlights.* https://population.un.org/wup/. Accessed on August 5, 2023.

Venkatarangan, L. B. (1989). Population Trends of First Four Megacities in India. In S. N. Singh et al. (Eds.), *Population Transition in India* (Vol. 2, pp. 309–318). B.R. Publishing Corporation.

Weeks, J. R. (2018). *Population: An Introduction to Concepts and Issues.* Rawat Publications, Jaipur.

# Chapter 7
# Urbanisation and Access to Basic Amenities

Urbanisation intrinsically entails an improvement in the access of basic amenities to the people. Basic amenities are essential for a good living condition and for the advancement of the social and economic health status. Studies show that basic amenities such as clean water, sanitation, electricity, clean fuel, etc., are critical determinants of health and well-being of people (Ali & Rahman 2004; Clegg & Garlick, 1979). Further, several studies show that urban areas have higher levels of household amenities compared to rural areas (Bhagat, 2013, 2017; Das & Mistry, 2013; Kumar, 2014). Thus, urbanisation plays an important role in the access to basic amenities. However, as urban areas are not homogeneous but vary in terms level of urbanisation and also size of urban centres and their municipal status, the level of basic amenities accessed by the urban population is likely to differ.

The bigger cities are known for agglomeration economies, having diversified economic structures and also higher income at the same time also have higher level of air pollution, crowding and also visible clusters of slums. It may be interesting to know how bigger cities differ in the access to basic amenities compared to medium and small urban centres. An understanding of the access to basic amenities by urban hierarchy is crucial for urban planning and urban governance as urban hierarchy continues to be a characteristic feature of urbanisation in most of the countries of the world, and same is applicable for India as well. This chapter presents how basic amenities like access to electricity, clean drinking water, toilet facility, and clean fuel like LPG vary by states and across the size class of towns and cities and their municipal status. This chapter also presents the situation in eight megacities namely Delhi, Mumbai, Kolkata, Chennai, Hyderabad, Bangalore, Pune and Ahmadabad which not only have very high concentration of population, but are also the engines of economic growth in the country.



R. B. Bhagat and M. I. Hassan, *Urbanisation and Urban Policies in India*, Exploring Urban Change in South Asia, https://doi.org/10.1007/978-981-96-3181-0_7

## 7.1 Census Data on Basic Amenities

As this chapter deals with the access to basic amenities in urban areas of states and also its variations by size and administrative status of urban centres, data required for the study is of aggregate and areal character. The census being a complete enumeration of households is the most appropriate data for the study as it is free from sampling errors. Census of India provides information on housing conditions, basic amenities and assets at the household level. Census in India is conducted in two phases, namely i). House listing and housing census; ii). Population enumeration. The data on housing conditions and basic amenities are collected in the house listing and housing census phase. The information related to housing conditions include building materials, usage of houses, condition of houses, ownership status of houses, housing stock, number of dwelling rooms, size of households, household by the number of married couples, etc. On the other hand, basic amenities of households relate to the source of drinking water, toilet, bathroom, drainage, cooking fuel, kitchen, source of lightening, etc. Further, information on household assets is also available such as radio, television, mobile and telephone, computer/laptop with and without internet connections and private modes of transport like bicycle, scooter/motorcycle/moped and car/jeep/van and the household availing banking facility. The information is available for rural and urban areas, for the state, town and city levels. Further, the data are also available for the administrative status of towns and cities such as Municipality and Municipal Corporation as well as for other non-municipal towns (Bhagat et al., 2018). Although presented census data are old but it can serve as a baseline for future evaluation of progress as it is based on a complete enumeration of households and also available for smaller administrative categories.

## 7.2 Basic Amenities

In this chapter, the following household characteristics and amenities are used to study the nature of urbanisation of states and size class of urban centres and their municipal status.

i) **Predominant Material of Roof:** A roof is an important part of a building. The predominant material of the roof is categorised in ten types and each one is given a separate code in the 2011 census. Out of ten types of roof materials concrete roof is one of the categories. The category of concrete roof type is comparatively more durable and reflects the condition of a house.
ii) **Main Source of Drinking Water:** Information on the main source of drinking water was classified into ten categories. In the 2011 census the tap water was bifurcated into two categories namely 'tap water from treated source' and 'tap water from untreated source'. In order to have a comparison with earlier census data in this study we have combined the both categories. The indicator 'drinking water' used in this chapter comprises of both tap water and also water from a

hand pump under the assumption both of them constitute as a source of clean drinking water.

iii) **Source of Lighting:** There are several sources of lighting in a household such as electricity, kerosene, solar, etc. Electricity as the main source of lighting is considered in this chapter.

iv) **Bathroom:** If a bathing facility is available in a household within the premises, the household was considered to possess a bathroom. Bathroom includes a bathroom or an enclosure without roof for the bathing facility within the premises. The latter was added in the 2011 census.

v) **Toilet Facility:** The census ascertained both availability and types of latrines in a household. When data is compared between censuses it is important to keep in mind different codes used by the respective censuses indicating the types of latrine facilities. We have considered only any latrine available in the household in this chapter. However, it would be interesting to look at the codes used in the censuses 2001 and 2011 as given below:

| Census 2001 | Code | Census 2011 | Code |
|---|---|---|---|
| No Latrine | 0 | **Flush/pour flush latrine connected to** | |
| Service Latrine | 1 | Piped water system | 1 |
| Pit Latrine | 2 | Septic system | 2 |
| Water Closet | 3 | Other system | 3 |
| | | **Pit Latrine** | |
| | | With slab/ventilated improved pit | 4 |
| | | Without slab/open pit | 5 |
| | | Night soil disposed into open drain | 6 |
| | | **Service Latrine** | |
| | | Night soil removed by human | 7 |
| | | Night soil serviced by animals | 8 |
| | | **No Latrine within premises** | |
| | | Public Latrine | 9 |
| | | Open | 0 |

*Source* Census of India (2011) *Instruction Manual for House listing and Housing Census*, Office of the Registrar General and Census Commissioner, India. Also see Bhagat et al. (2018)

vi) **Type of Fuel Used for Cooking:** Types of fuel used for cooking such as firewood, crop residue, cow dung cake, electricity, biogas, LPG/PNG, etc., were ascertained in the census. If the household is using more than one fuel for cooking, the predominant fuel used for cooking was recorded. Only LPG/PNG as a source of fuel used for cooking is considered in this study.

vii) **Drainage System:** The system of disposal of wastewater and other liquid wastes from a house is recognised as a drainage system. The drainage system drains out wastewater and other liquid wastes emanating from a household.

It may consist of kitchen wastewater, bath and wash water. In the census, it was ascertained whether it was a closed, open drainage or no drainage that connects the household. In this chapter, we have considered a household with drainage if it is connected with any drainage system.

viii) **Availability of Kitchen:** Where meals are cooked is an important marker of the status of a household. It may be cooked inside the house in a separate room or outside with no room designated for it. If a household cooked meals in a separate room meant for it, it was considered to have a kitchen available in the household.

## 7.3 Access to Basic Amenities: National and State-Level Patterns

Figure 7.1 presents the percentage of households with basic amenities available such as houses with concrete roof, bathroom and kitchen, access to clean drinking water, electricity, toilets, drainage and LPG/PNG in India for the combined rural and urban areas. About 30 per cent of total households had a concrete roof in 2011—an increase of only 9 per cent from 2001.

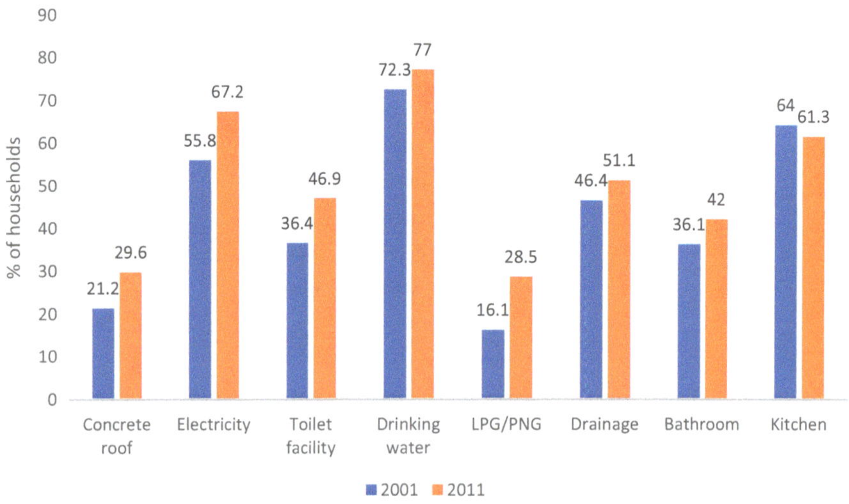


**Fig. 7.1** Percentage of households with selected basic amenities, India, Total, 2001–2011 (*Source* Drinking Water—Tap water and hand pump; Bathroom includes i) bathroom within premises and ii) an enclosure without roof as per the 2011 census, however, the latter information was not collected in the 2001 census. In order to make comparable only bathroom within the premises is presented in this graph. LPG also included PNG. *Source* Census of India 2001 and 2011, House listing and Housing Census data, H Series Tables on Houses, Household Amenities and Assets Data, https://censusindia.gov.in/census.website/data/census-tables#)

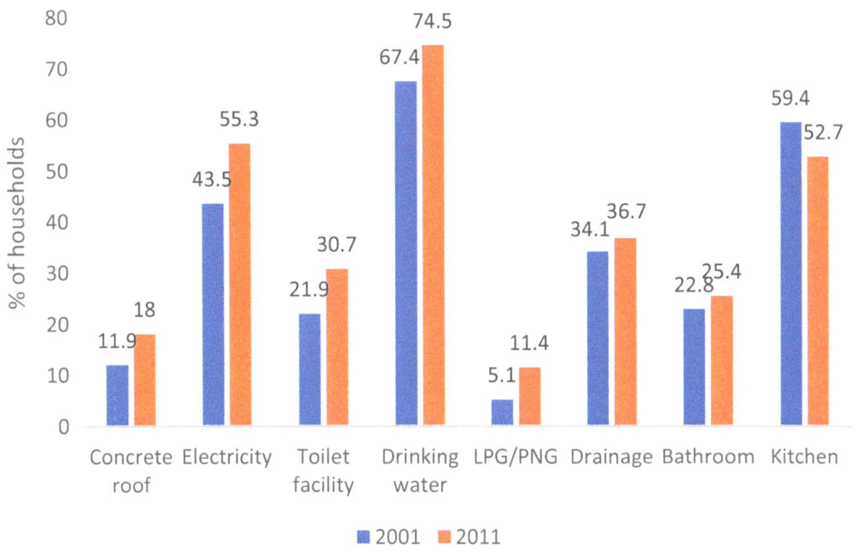


**Fig. 7.2** Percentage of households with selected basic amenities, India, Rural, 2001–2011 (*Source* Same as Fig. 7.1)

In urban areas, there were about three times higher houses with concrete roofs compared to rural areas in 2011. In 2011, there were 67 per cent of total households have electricity compared to 56 per cent in 2001. On the other hand, the percentage of households with access to electricity was about 93 per cent in urban areas compared to 55 per cent in rural areas. In 2011 about 47 per cent of total households have any toilet facility and it varied from about 31 in rural to 81 per cent in urban areas. While 80 per cent of urban households have access to clean drinking water in 2011, there has been a decline in coverage of households with clean drinking water during 2001–2011. About 82 per cent of urban households were connected to any drainage facility and 81 per cent of households of any toilet facility in urban areas in 2011 (see Fig. 7.3).

In 2011, there were only about 29 per cent households using LPG/PNG as cooking fuel compared to just 16 per cent of all households in 2001. In urban areas over three-fifth used LPG/PNG compared to one-tenth of the households in rural areas until 2011. Only half of the total households were covered with any drainage facility in 2011—a little increase from 2001 (46 per cent). On the other hand, there was a huge difference between rural (37 per cent) and urban areas (82 per cent) in 2011 at the all-India level. Figure 7.3 clearly shows that all urban households do not have sanitation facilities as late as the end of the first decade of twenty-first century.

The ease of living in urban areas depends upon the access to electricity, toilet facility, clean water, fuel and drainage along with the availability of open spaces, playgrounds and parks. It is generally believed that the availability of basic amenities is higher in urban compared to rural areas. Water and sanitation are the most

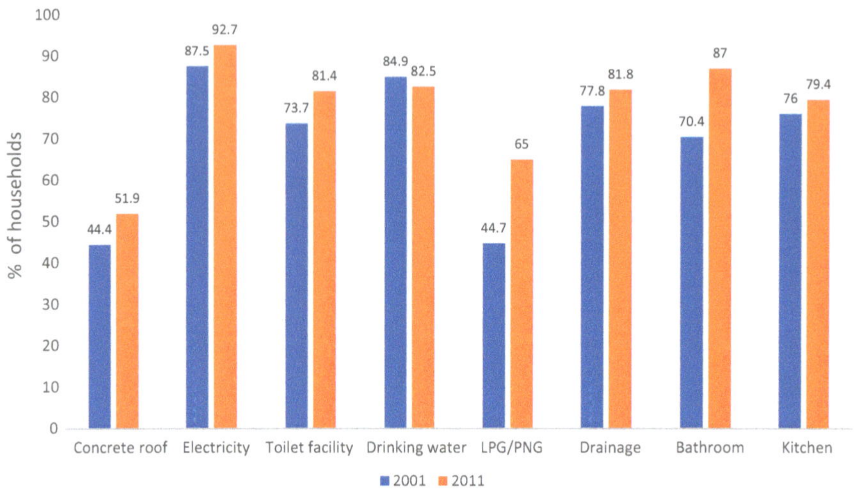


**Fig. 7.3** Percentage of households with selected basic amenities, India, Urban, 2001–2011 (*Source* Same as Fig. 7.1)

important components of basic amenities closely connected with not only economic status but the social status as well. All the components of basic amenities are in fact inter-connected and cannot be functional independent of each other. For example, sanitation is not only an absence of garbage and waste materials strewn around but also inextricably related to the access to toilet facility, water and connectivity to a drainage system. The lack of sanitation facilities for a huge number of households was recognised by the erstwhile Planning Commission (Planning Commission, 2008). In 2008, the Ministry of Urban Development adopted an Urban Sanitation Policy with a focus on hygienic and affordable sanitation facilities for the urban poor and women. Subsequently, the efforts to achieve universal sanitation coverage were augmented with the launch of *Swachh Bharat Mission* (Clean India Mission) in both rural and urban areas on 2 October 2014—i.e. the birthday of the father of the nation Mahatma Gandhi. Although the Clean India Movement or *Swachchata Andolan* initiated was a noble initiative, merely cleaning of streets and government offices may not guarantee basic amenities as a matter of right to life under Article 21 of the Indian Constitution.

The preparation of citizen's charters could be a very important step to implement and ensure access to basic amenities. It is also important to improve urban governance as the provision of basic amenities is largely the part of public policy. In order to achieve sustainable urbanisation and meet the sustainable development goals there is a need to have a vigorous planning and promotion of recycling of wastes—both solids and liquids and advance a circular economy not only environmentally friendly but also capable of generating more jobs. In the circular economy, water and sanitation assume utmost significance. It is about waste collection, segregation recycling and management. It is about how we keep the sources of water clean by sewage treatment and also about recycling of wastewater for industrial and agricultural use. It requires not only an integrated plan but also a huge investment as well. As a follow-up of the implementation of Clean India Movement, the government should continue to innovate an integrated action plan and solution for the sustainable sanitation and adequate clean water for each urban centres and villages in India (Bhagat, 2014).

However, it is important to emphasise that for a country like India, one size fits all may not be desirable. Moreover, sanitation and water are a local issue and a decentralised approach to their augmentation and management may be more pertinent. In both rural and urban areas, the local governments must lead and made accountable and with adequate empowerment, resources and a sanitation and water strategy that is just, inclusive and sustainable.

## 7.4 State Level Patterns

India manifests a huge variation in economic and social conditions closely related to the pattern of urbanisation. The central and eastern states along with north-eastern state of Assam showed a very low level of urbanisation—lower than the national average. On the other hand, urbanisation level is higher in southern, western and northern India. It is expected that states with lower levels of urbanisation will also show lower levels of basic amenities accessible to the households. Table 7.1 presents the percentage of households availing basic amenities for urban India for the census year 2011. While 52 per cent of the households had a concrete roof in urban India, it varied from 11 per cent in Manipur to 83 per cent in Chandigarh. The mountainous states such as Uttarakhand (58 per cent), Himachal Pradesh (64.6 per cent) and Sikkim (55 per cent) had a higher percentage of houses with a concrete roof than the national average (52 per cent). This indicates why these states were vulnerable to natural disaster in the recent past. It is known that the use of locally available materials in the construction of houses declines as a state or region urbanises (Bhagat & Mishra, 1989).

Among states, the availability of electricity was one of the lowest in Bihar (66 per cent). Some of the lagging states like Jharkhand, Uttar Pradesh and Assam have much lower access to electricity at the household level compared to the national average (about 93 per cent). Table 7.1 also shows Chhattisgarh (60 per cent) had one of the lowest percentage of households covered by toilet facility followed by Orissa (65 per cent) and Jharkhand (67 per cent). On the other hand, households in almost all north-eastern states were found to have higher toilet facility compared to the national average (81 per cent), however, same was not true for drinking water supply. Earlier studies showed that pit latrines were more prevalent in north-eastern states (Bhagat, 2011). Further, in north-eastern states, a large number of households have been depending upon streams and rivulets for water. On the other hand, in Kerala a significant percentage of households use well water for drinking purposes (Kundu, 1999).

The access to clean fuel like LPG/PNG varied from a lowest of 42 per cent in Jharkhand followed by Chhattisgarh (45 per cent) and Odisha (46 per cent) among

**Table 7.1** Percentage of households with selected basic amenities in urban areas by states and UTs, India, 2011

| State/UTs | Concrete Roof | Drinking Water | Electricity | Toilet | Drainage | Bathroom | LPG/PNG | Kitchen | Index |
|---|---|---|---|---|---|---|---|---|---|
| India | 51.9 | 82.5 | 92.7 | 81.4 | 81.8 | 87.0 | 65.0 | 79.4 | 0.7086 |
| Andaman & Nicobar Islands | 23.6 | 98.0 | 97.7 | 87.1 | 86.0 | 93.9 | 66.9 | 95.0 | 0.7884 |
| Andhra Pradesh | 64.6 | 87.9 | 97.3 | 86.1 | 88.3 | 92.6 | 67.7 | 77.0 | 0.8057 |
| Arunachal Pradesh | 12.1 | 89.8 | 96.0 | 89.5 | 66.2 | 80.6 | 77.5 | 93.0 | 0.6826 |
| Assam | 15.1 | 63.1 | 84.1 | 93.7 | 56.3 | 84.5 | 68.0 | 91.0 | 0.5806 |
| Bihar | 51.9 | 84.3 | 66.7 | 69.0 | 71.4 | 68.6 | 47.4 | 53.0 | 0.4135 |
| Chandigarh# | 83.0 | 98.6 | 98.4 | 87.6 | 96.0 | 89.7 | 71.8 | 74.0 | 0.8675 |
| Chhattisgarh | 44.5 | 80.6 | 93.7 | 60.2 | 68.9 | 67.4 | 45.0 | 73.0 | 0.4901 |
| Dadar & Nagar Haveli# | 41.7 | 69.5 | 98.5 | 81.3 | 62.7 | 90.0 | 66.2 | 87.0 | 0.6805 |
| Daman & Diu# | 37.5 | 78.0 | 99.3 | 85.4 | 77.2 | 84.2 | 55.9 | 62.0 | 0.6574 |
| Goa | 41.5 | 90.2 | 97.7 | 85.3 | 78.0 | 94.9 | 81.1 | 94.0 | 0.8110 |
| Gujarat | 67.6 | 88.9 | 97.2 | 87.7 | 82.6 | 90.5 | 68.3 | 83.0 | 0.8048 |
| Haryana | 57.3 | 85.9 | 96.2 | 89.9 | 92.2 | 92.6 | 77.6 | 80.0 | 0.8283 |
| Himachal Pradesh | 64.6 | 96.9 | 98.1 | 89.1 | 93.6 | 92.8 | 85.1 | 91.0 | 0.8951 |
| Jammu & Kashmir | 39.2 | 94.4 | 98.0 | 87.5 | 83.1 | 91.4 | 75.7 | 94.0 | 0.8076 |
| Jharkhand | 54.0 | 68.7 | 88.0 | 67.2 | 70.8 | 69.8 | 42.1 | 60.0 | 0.4717 |
| Karnataka | 55.9 | 81.7 | 96.4 | 84.9 | 87.6 | 96.4 | 64.3 | 95.0 | 0.8052 |
| Kerala | 58.0 | 35.5 | 97.0 | 97.4 | 54.5 | 92.9 | 48.4 | 98.0 | 0.6589 |
| Lakshadweep# | 47.9 | 20.1 | 99.7 | 97.7 | 25.9 | 97.6 | 20.9 | 96.0 | 0.5353 |
| Madhya Pradesh | 44.5 | 76.8 | 92.7 | 74.2 | 82.3 | 85.3 | 60.5 | 70.0 | 0.6425 |
| Maharashtra | 52.0 | 92.0 | 96.2 | 71.3 | 91.2 | 95.4 | 74.1 | 82.0 | 0.7852 |
| Manipur | 10.8 | 60.5 | 82.4 | 95.8 | 70.4 | 65.7 | 56.4 | 6.0 | 0.3925 |

(continued)

**Table 7.1** (continued)

| State/UTs | Concrete Roof | Drinking Water | Electricity | Toilet | Drainage | Bathroom | LPG/PNG | Kitchen | Index |
|---|---|---|---|---|---|---|---|---|---|
| Meghalaya | 32.8 | 78.0 | 94.9 | 95.7 | 80.3 | 79.9 | 45.6 | 95.0 | 0.6713 |
| Mizoram | 24.7 | 75.2 | 98.1 | 98.5 | 79.5 | 89.7 | 83.8 | 93.0 | 0.7843 |
| Nagaland | 15.7 | 39.7 | 97.4 | 94.6 | 76.5 | 94.4 | 52.7 | 96.0 | 0.6532 |
| NCT of Delhi# | 64.6 | 87.0 | 99.1 | 89.8 | 96.0 | 91.3 | 90.3 | 79.0 | 0.8779 |
| Odisha | 46.9 | 60.8 | 83.1 | 64.8 | 59.0 | 64.3 | 46.1 | 74.0 | 0.4016 |
| Puducherry# | 70.9 | 95.9 | 98.5 | 82.0 | 82.6 | 92.0 | 78.7 | 84.0 | 0.8387 |
| Punjab | 72.0 | 86.7 | 98.3 | 93.4 | 90.9 | 94.8 | 79.3 | 83.0 | 0.8825 |
| Rajasthan | 28.4 | 88.7 | 93.9 | 82.0 | 86.0 | 89.2 | 69.4 | 76.0 | 0.7128 |
| Sikkim | 55.0 | 92.1 | 98.7 | 95.2 | 92.1 | 92.0 | 85.1 | 90.0 | 0.8843 |
| Tamil Nadu | 56.7 | 84.3 | 96.1 | 75.1 | 74.9 | 85.2 | 68.7 | 89.0 | 0.7139 |
| Tripura | 13.5 | 75.4 | 91.6 | 97.9 | 53.4 | 64.0 | 49.7 | 92.0 | 0.5182 |
| Uttar Pradesh | 30.7 | 88.3 | 81.4 | 83.1 | 93.4 | 84.6 | 61.7 | 68.0 | 0.6402 |
| Uttarakhand | 58.0 | 95.6 | 96.5 | 93.6 | 92.9 | 93.1 | 79.4 | 84.0 | 0.8690 |
| West Bengal | 48.4 | 78.2 | 85.1 | 85.0 | 66.8 | 70.7 | 48.2 | 68.0 | 0.5341 |

*Source* Census of India 2001 and 2011, House listing and Housing Census data, H Series Tables on Houses, Household Amenities and Assets Data, https://censusindia.gov.in/census.website/data/census-tables#

The information for bathroom includes both bathroom within premises as well as an enclosure without roof major states compared to more than 80 per cent households had LPG/PNG in Goa, Himachal Pradesh, Mizoram, Sikkim and Delhi in 2011. At the state level, the level of basic amenities closely follows the level of urbanisation. It means that the states with low availability of electricity in urban areas were also the states with low levels of urbanisation. A correlation analysis showed a positive significant correlation co-efficient ($r = +0.53$) between the proportion of households with electricity in urban areas and the proportion of population living in urban areas. Similarly, the percentage of urban is also significantly positively correlated with LPG/PNG (+0.35) at the state level (Bhagat, 2011). On the other hand, the availability of toilet, drinking water and drainage were not significantly correlated. Earlier studies had also shown that the provision of sanitation facility did not reflect improvement with increasing economic status at the household level (Kundu, 1991). Housing characteristics such as households with concrete roofs, bathroom and kitchen also depict positive but low correlation. Overall, the amenities index of the combined effects of all indicators is also presented in Table 7.1. The household amenities in urban areas showed that Himachal Pradesh, Sikkim, Punjab, Delhi and Uttarakhand were the top five states, whereas Manipur, Odisha, Bihar, Jharkhand and Chhatisgarh were the bottom five states according to the 2011 census.

Overall the household amenity index showed a positive and significant relationship with the level of urbanisation at the state level ($r = +0.44$) (See Fig. 7.4).

## 7.5 Access to Basic Amenities by Size Class of Cities/Towns

While India's urban population is distributed across 8000 odd towns and cities with different sizes, the economic base and ability to generate tax and non-tax resources vary widely. It is generally observed that cities (100 thousand and more) have higher employment opportunities compared to towns. In many smaller urban centres, a sizeable proportion of workforce is also engaged in agriculture. As such, the urban size reflects not only population concentration but also their economic strength. It is likely that the level of basic amenities is directly related to the size of urban centres. Table 7.2 which presents basic amenities by size class of urban centres confirms this conjecture except for toilet, bathroom, kitchen and drainage facilities which was found relatively in lower proportion compared to cities in general. It may be noted that the toilet facility was not only low (less than 60 per cent) among small urban centres (less than 50 thousand), but also in megacities with more than 5 million population (58 per cent). Megacities are also known to have a higher proportion of population living in slum areas that have either no access to toilet facility or having a community toilet (Bhagat, 2011). It is very important to mention here that the response on toilet facility relates to the latrine facility available within the premises, and it does not include either a public or a community toilet. However, the condition in regard to electricity and supply of drinking water is much better in megacities (population 5 million and more) with 98 per cent of households reported to have the access compared to more than 80 per cent in the cities. On the other hand, households

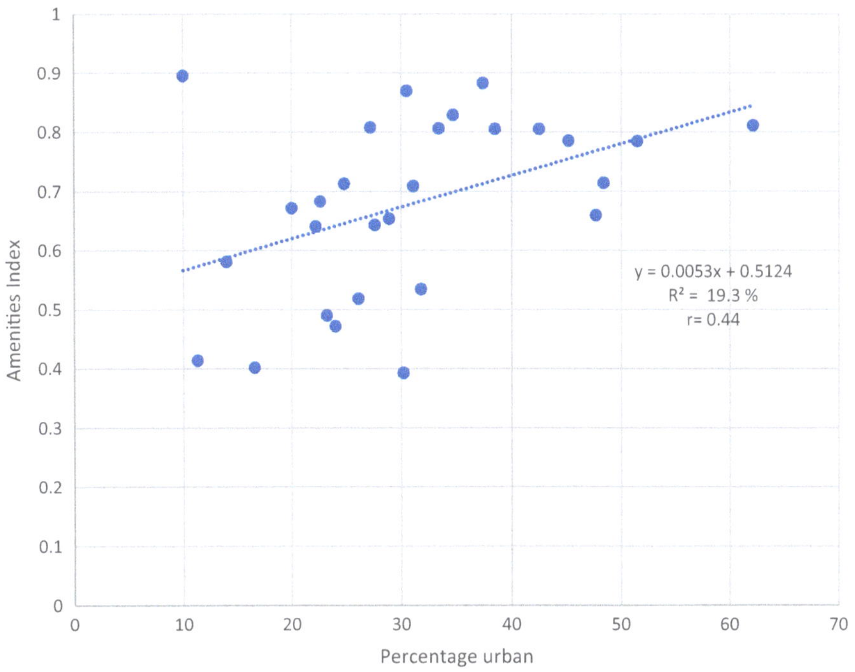


**Fig. 7.4** Level of urbanisation and access to basic amenities in urban areas, state level pattern, India, 2011

living in small urban centres show relatively lower level of these amenities. For example, two-third of households have access of LPG/PNG in megacities compared to one-fourth in the small towns (less than 10,000) (see Table 7.2). However, it is also important to mention that a significant proportion of households living in urban areas have been using kerosene coal, charcoal and wood as sources of fuel which are sources of indoor pollution and ill health (Bhagat, 2011).

Even with different levels of urbanisation, the bigger urban centres in their respective states show a higher level of provision of the basic amenities compared to smaller urban centres. However, the inter-state variations continue to reflect the disparities by size class of urban centres. It means that the bigger urban centres of less developed states like Bihar, Orissa, Jharkhand and Uttar Pradesh show much lower level of basic amenities compared to the same categories of urban centres of Punjab, Maharashtra, Gujarat and Karnataka. Hence, the inter-state disparities continue to be apparent within the same size class category (Bhagat, 2011).

The size of an urban centre is closely associated with its economic base, revenue potential, diversification of the economy and the structure of municipal governance. In India, the state government is mandated to decide the structure of municipal governance by enacting a municipal act as urban development is a state subject. In general, bigger cities are governed by Municipal Corporations, while various forms of municipalities govern the smaller cities. The various forms are named such as

**Table 7.2** Percentage of households with selected basic amenities by size class of urban centres, India, 2011

| India | Electricity Available | Toilet facility Available | Drinking water Tap + Handpump | Bathroom facility within the house | Separate kitchen within the house | LPG/ PNG | Drainage Close + open |
|---|---|---|---|---|---|---|---|
| More than 5 million | 97.21 | 57.67 | 97.52 | 74.28 | 70.55 | 63.02 | 82.80 |
| 5 million–1 million | 86.69 | 78.49 | 89.75 | 83.37 | 81.01 | 59.92 | 90.09 |
| 1 million–1 lakh | 80.99 | 72.93 | 85.06 | 73.22 | 77.20 | 50.50 | 78.43 |
| Class II (50–100 thousand) | 77.70 | 66.40 | 81.82 | 67.54 | 74.81 | 43.71 | 73.31 |
| Class III (20–50 thousand) | 76.56 | 62.49 | 78.25 | 60.41 | 72.99 | 35.55 | 67.31 |
| Class IV (10–20 thousand) | 78.28 | 57.39 | 78.89 | 55.06 | 71.71 | 29.76 | 63.96 |
| Class V (5–10 thousand) | 76.27 | 53.86 | 78.61 | 49.76 | 70.74 | 26.42 | 57.96 |
| Class VI (Less than 5 thousand) | 77.93 | 62.53 | 71.29 | 40.57 | 56.99 | 26.59 | 50.82 |

*Source* Same as (Table 7.1)

the municipal council, municipal committee, municipal board, etc. The Cantonment Board is usually established by the Ministry of Defence for civil population living in the cantonment area. In addition to these, there are many structures of urban local bodies like Notified Area, Notified Area Committee, Town Committee, Township, etc.

However, it is important to note that non-municipal towns or census towns are governed mostly by Gram Panchayats even though they are declared as urban centres. Table 7.3 presents the percentage of households availing different basic amenities by municipal status. It may be seen from the table that municipal structure in India is very diverse. In general, Table 7.3 shows that municipal towns are better in terms of provision of basic amenities compared to census towns. This indicates that municipalisation of census towns would be helpful in improving the conditions of census towns which are mostly governed by Gram Panchayats. It may also be noted that Cantonment Board, Industrial Township and Notified Towns are showing better level of basic amenities. They are also governed by different administrative entities and generally smaller in size. It is possible that their financial resources other than local tax and revenue may have some impact. It is also important to mention that household toilets, LPG/PNG, bathrooms and kitchens fall under the category of private goods whereas electricity, drinking water and drainage facilities also depend upon the provision of public goods provided in the respective urban centres. Looking from this perspective, Table 7.3 shows that Municipal Corporation with large population, tax and revenue bases have generally higher level of basic amenities. It is thus very important to recognise the economies of scale in urban policies and programmes.

**Table 7.3** Percentage of households with selected basic amenities by municipal status, India, 2011

| Municipal Status | No. of Towns | Drinking Water | Electricity | Toilet | Drainage | Bathroom | LPG/ PNG | Kitchen |
|---|---|---|---|---|---|---|---|---|
| Municipal Corporation | **218** | 87.5 | 96.6 | 86.4 | 92.1 | 93.6 | 74.4 | 83.1 |
| Cantonment Board | **58** | 89.4 | 95.7 | 78.4 | 91.9 | 89.9 | 75.9 | 78.4 |
| City Municipal Council | **44** | 81.4 | 95.3 | 76.2 | 83.8 | 94.7 | 53.8 | 92.2 |
| Census Town | **3865** | 65.4 | 89.9 | 79.3 | 62.1 | 79.7 | 51.2 | 77.5 |
| Industrial Notified Area | **26** | 87.4 | 86.7 | 83.4 | 79.6 | 85.7 | 70.0 | 76.5 |
| Industrial Township | **10** | 82.7 | 91.8 | 75.1 | 81.3 | 81.2 | 62.7 | 80.7 |
| Municipality | **916** | 79.7 | 93.2 | 80.3 | 75.1 | 83.4 | 61.1 | 77.6 |
| Municipal Council | **415** | 86.9 | 95.0 | 80.9 | 87.5 | 92.7 | 71.9 | 81.8 |
| Municipal Board | **37** | 72.1 | 88.4 | 95.4 | 64.4 | 87.7 | 70.3 | 91.0 |
| Municipal Committee | **64** | 86.2 | 97.6 | 87.6 | 93.9 | 89.9 | 84.1 | 81.5 |
| Notified Area | **7** | 91.7 | 90.0 | 84.9 | 67.8 | 78.6 | 64.5 | 79.1 |
| Notified Area Committee | **72** | 70.7 | 82.7 | 64.3 | 62.8 | 63.9 | 46.3 | 72.7 |
| Nagar Panchayat | **1099** | 83.4 | 69.5 | 61.2 | 72.8 | 67.2 | 37.5 | 54.0 |
| Nagar Palika Parishad | **223** | 90.6 | 81.4 | 85.5 | 94.1 | 84.6 | 59.9 | 67.6 |
| Notified Town | **48** | 80.4 | 97.4 | 95.4 | 74.7 | 86.6 | 81.7 | 89.4 |

(continued)

(continued)

| Municipal Status | No. of Towns | Drinking Water | Electricity | Toilet | Drainage | Bathroom | LPG/ PNG | Kitchen |
|---|---|---|---|---|---|---|---|---|
| Town Committee | **75** | 60.6 | 83.3 | 91.9 | 43.3 | 79.0 | 48.8 | 93.1 |
| Town Municipal Council | **94** | 82.0 | 92.9 | 65.5 | 77.5 | 92.4 | 46.1 | 91.6 |
| Town Panchayat | **629** | 81.5 | 92.7 | 58.8 | 61.3 | 75.0 | 54.6 | 84.1 |
| **Total** | **7900** | **82.0** | **92.6** | **81.3** | **81.7** | **86.9** | **65.0** | **79.3** |

*Source* Same as (Table 7.1)

## 7.6 Megacities

Megacities are the dominant urban spaces. Among the eight megacities UA namely Delhi, Mumbai, Kolkata, Chennai, Hyderabad, Bangalore, Pune and Ahmadabad the toilet facility was available to a maximum of 98 per cent households in Hyderabad followed by Bangalore (97 per cent) and lowest in Mumbai (64 per cent) in 2011. In Pune 81 per cent and in Delhi 90 per cent of the households had toilet facilities in 2011. It is known that some cities such as Mumbai and Delhi have huge slum population, the access to toilet facility was lower due to the fact that toilet facility refers to its availability within the premises, and does not include community and public toilets used by slum populations. Providing toilet facility is a great challenge for megacities with huge slum populations even today. The availability of clean water is a grave problem not only for drinking purpose but also in sustaining community and public toilets in slum areas. It also poses a big challenge for the communities, NGOs and urban local bodies in providing toilet and sewer facilities in slums along with water supply (Bhagat, 2014). In addition, the problem of sanitation is aggravated by the lack of garbage collection system in many parts of urban areas including mega urban areas (Patel, 2013). Further, in several slum areas there exists an open drainage which is required to be cleaned regularly to prevent the choking of sewers as many people often throw garbage in the open drainage. Thus, overflowing of open drainage and water logging are common problems in many cities and towns not only confined to slum areas but also in other areas as well. It not only affects quality of life but also incurs a huge health risk. A proper drainage system, sewage treatment and recycling of wastewater constituting an integrated sanitation strategy are often lacking in the majority of the urban centres. It is very important to underscore that sanitation is a larger issue and a holistic strategy is most desirable for the urban areas in general and slum areas in particular (Bhagat, 2014).

In some of the megacities water crisis has been acute. For example, about one-fourth of households had no access to drinking water in Chennai UA followed by

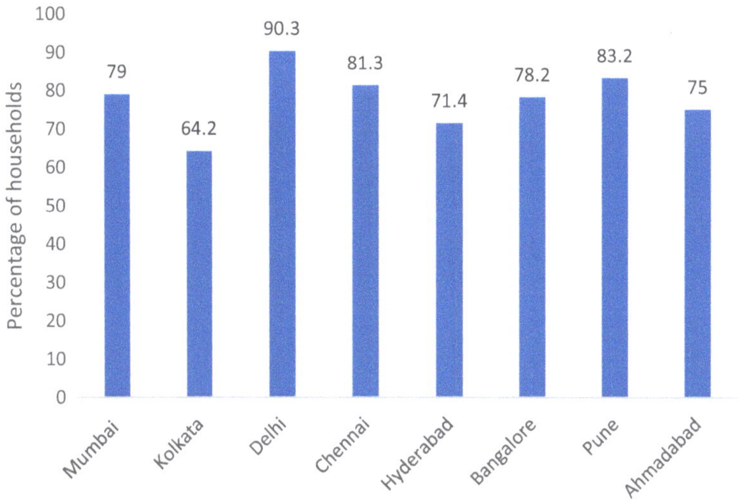


**Fig. 7.5** Percentage of households availing LPG/PNG in megacities UA, 2011 (*Source* Same as Table 7.1)

one-fifth in Bangalore UA, while most of the households had access to drinking water in Mumbai, Hyderabad and Pune UAs. In Chennai and Ahmadabad, Kolkata drainage facility was lacking to some extent compared to other megacities where at least one-tenth of households had no access to either closed or open drainage for the outlet of wastewater from the households as per 2011 census (Bhagat et al., 2018). The use of LPG/PNG varied from nearly 64 per cent of households in Kolkata to 90 per cent in Delhi UA followed by about 80 per cent in Pune and Chennai (Fig. 7.5). In several of the megacities a sizeable proportion of households also use kerosene as household fuel (Bhagat, 2011).

The mega urban UAs are known for their contrasts between the core and periphery of the mega UA. Studies also show that peripheral areas of megacities have grown faster than the core (Sita & Bhagat, 2007). If we define Municipal Corporation areas as core and rest of the UA as periphery which consists of many small and medium statutory towns as well as census towns, the contrast in the provision of amenities is very much visible with respect to provision of water and sanitation. In general, peripheries of megacities are lacking in the provision of water supply particularly piped water supply and drainage facilities. The situation is mixed so far as the provision of other basic amenities is concerned. For example, Fig. 7.6 presents the data on LPG/PNG by the core and the peripheries of megacities which vindicate that the core and periphery contrasts are not uniform and could not be generalised for all megacities UA.

In the changed scenario of development showing the presence of industrial estates, housing enclaves, educational institutions, malls, etc., along the highways connecting the core of megacities, it may be pertinent to question the concept of degenerated peripheralisation advanced almost a decade ago (Kundu, 2006). However, it is not

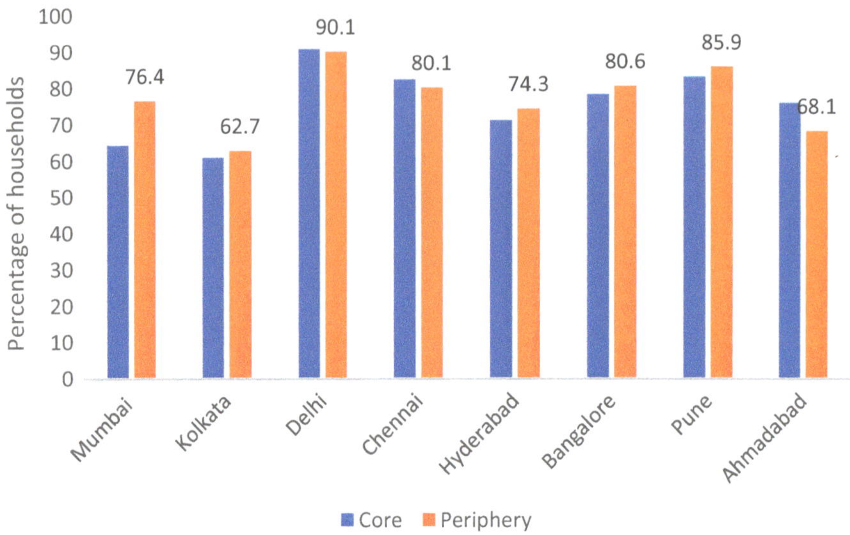


**Fig. 7.6** Percentage of households availing LPG/PNG in the core and periphery of megacities UA, 2011 (*Source* Same as Table 7.1)

to argue that the peripheries have not been degraded in terms of intensive land use, depletion of forests and exploitation of water resources but the forces of market and capital accumulation have been widened their footprints in the extended mega urban agglomerations.

## 7.7 Conclusions and Suggestions

The level of urbanisation is significantly positively associated with higher access to household amenities. A composite index constructed based on the selected indicators of concrete roof, electricity, drinking water, drainage facility, LPG/PNG and availability of bathroom and kitchen facilities shows that the urban areas of the Himachal Pradesh, Sikkim, Punjab, Delhi and Uttarkhand are the top five states, whereas Manipur, Odisha, Bihar, Jharkhand and Chhatisgarh are the bottom five states in access to household amenities. Hilly states like Himachal Pradesh and Sikkim although do not have higher level of urbanisation as per conventional measure of urban definition, the provision of basic household amenities reflects higher access. However, the same situation is not found in north-eastern states which are also located in hilly terrain. Thus, government policies and programmes do matter, and urban governance plays an important role.

The proportion of households having access to basic amenities also varies by size class of urban centres. Cities and megacities generally demonstrate higher proportion of households served by basic amenities compared to small and medium urban centres. While the basic amenities like electricity, LPG and drinking water show a direct relationship with the size class of cities and towns, this, however, is not true for the amenities like toilet facility as some of the megacities show lower proportion of households with toilets within premises due to higher presence of slum households. The existing pattern in the distribution of basic amenities at the national level by size class of cities and towns is also found true at the state level. Thus, it reveals that bigger cities and towns of high-income states like Punjab, Gujarat, Karnataka, Tamil Nadu and Maharashtra show higher proportion of households with basic amenities compared to the same categories of cities and towns in low-income states like Bihar, Orissa, Chhattisgarh, Madhya Pradesh, Rajasthan and Jharkhand.

On the whole, the analysis of data on basic amenities very importantly reveals the persistence of regional disparities both at the urban and the city/town levels. While the regional disparities at the inter-city level require a robust spatial strategy at the state level, intra-city inequalities need to be addressed through improving urban policy, planning and programmes. As we know that the 74th Amendment to the constitution mandates to the urban local bodies to take up several areas of urban planning and development including public health, sanitation, solid waste management, etc. (Chakrabarti, 2001). However, the empowerment and autonomy of the urban local bodies as mandated in the 74th Amendment to the constitution which came into effect since 1992 is still lacking its full implementation. While the state governments are ambivalent and lack commitment to a decentralised governance, the flagship programmes of central government such as Jawaharlal Nehru National Urban Renewal Mission (JNNURM) earlier, and Smart Cities Mission and AMRUT (Atal Mission for Rejuvenation and Urban Transformation) later act as a centralised model of urban development making state governments and local bodies dependent on the centre. It also promotes 'one size fits all' programmes and hampers the progress in the decentralised governance, planning and urban development.

Since the liberalisation of the Indian economy in the 1990s, the urban local bodies are also expected to raise their own resources to meet the growing demand for basic amenities in urban areas. Over time there has been a rising expenditure not only on governance but also the need for the investment in infrastructure projects on water, sanitation, health, recreation, transport, etc., have also been growing. However, many urban centres particularly the small ones have little financial capacity and also lack technical capabilities to design projects and raise funds from the market. The state governments have not suitably empowered them to take up urban planning and governance independently including the power to raise money through taxation and market (Bhagat, 2011). On the other hand, several state governments have abolished octri—a major source of income to the urban local bodies even before the Goods and Service Tax (GST) came into existence, which amalgamated octri into a single tax known as GST since 1 July 2017 (Bhagat, 2005). Added to the low capacity and financial status of many urban centres, the centrally sponsored urban development policies programmes are designed in such a manner and spirit that embodies competitiveness, capacity and efficiency making them biased towards the metropolises and megacities. On the other hand, many small and medium urban centres have not been able to avail the opportunities of central government programmes due to their low technical and planning capacities, although they play an important role in serving the rural areas as a service centre. Their development is crucial for the rural development and urban transformation of rural areas. Further, as mentioned earlier the provision of basic amenities is low in the small urban centres, and their population has been growing both due to natural increase and migration. The basic amenities are not only low but also unable to meet the challenges of even a moderate growth of population in small urban centres. In fact, the challenges of providing basic amenities in general and water and sanitation and clean fuel in particular among all small urban centres are enormous as they lack resources of their own and are not able to attract investments from the private sources. SDG Goal 6 emphasises a universal access to water, sanitation and hygiene as these are the most basic human needs for life, health and well-being and efforts should be made to meet the requirements by 2030. It is also emphasised in the UN reports that the overall target to expand the access to clean water and sanitation in urban areas will not be met unless there is a significant improvement in the provision of services in small urban centres (UN-Habitat, 2006). The problem is no less important for some of bigger cities harbouring with a large slum population. In addition to public provisions of basic amenities, a proper maintenance and augmentation of basic amenities at the community level is extremely needed for their sustainability. An adequate provision of clean water, proper sanitation and the proper outlet of wastewater and its treatment would not only prevent occurrences of many infectious and parasitic diseases improving the quality of in the urban areas, but also increase productivity, efficiency and ensure sustainability in the long run.

**Acknowledgements** This chapter is a revised version of the article of the first author appeared as Bhagat, R.B. (2011) 'Urbanisation and Access to Basic Amenities in India', *Urban India,* 31(1):1-13. We are thankful to the Journal for permission. It also partly draws on the subsequent work of the first author namely Bhagat, R.B., Sarode, Sunil and Dwivedi, L.K., Kumar, Mausam and Lokhande, Sachin (2018) 'Access to Household Amenities and Assets in India: A Census Based Study', Research Monograph, International Institute for Population Sciences, Mumbai (https://iipsindia.ac.in/sites/default/files/Report_Asset_Amenities_Bhagat2.pdf).

## References

Ali, Osman, Z. M., & Rahman, M. R. A. (2004). The Effect of Urbanisation on the Health of Urban Residents. *Akademika, 65*, 111–124.

Bhagat, R. B. (2005). Rural-Urban Classification and Municipal Governance in India. *Singapore Journal of Tropical Geography, 26*, 61–74.

Bhagat, R. B., & Mishra, U. S. (1989). Types of Dwellings in Relation to Environment: A Case Study of Orissa. *Journal of National Building Organisation, 34*(2), 21–24.

Bhagat, R. B. (2017). Knowing the Social World Through Census: Reflections on the Conditions of Scheduled Castes and Scheduled Tribes. In N. Jayaram (Ed.), *Knowing the Social World: Perspectives and Possibilitie*s (pp. 23–38). Orient Black Swan.

Bhagat, R. B. (2011). Urbanisation and Access to Basic Amenities in India. *Urban India, 31*(1), 1–13.

Bhagat, R. B. (2013). Conditions of SC/ST Households. *Economic & Political Weekly, 48*(41), 63.
Bhagat, R. B. (2014, December). Rural and Urban Sanitation in India. *Kurushetra*, 11–14.
Bhagat, R. B., Sarode, S., Dwivedi, L. K., Kumar, M. & Lokhande, S. (2018). *Access to Household Amenities and Assets in India: A Census Based Study, Research Monograph, International Institute for Population Sciences, Mumbai.* https://iipsindia.ac.in/sites/default/files/Report_Asset_Amenities_Bhagat2.pdf.
Census of India. (1991). *Emerging Trends of Urbanisation in India.* Occasional Paper No. 1 of 1993, New Delhi: Registrar General, India.
Chakrabarti, P. G. D. (2001). Urban Crisis in India: New Initiatives for Sustainable Cities. *Development and Practice, 2&3*, 260–272.
Clegg, E. J., & Garlick, J. P. (1979). The Ecology of Disease in Urban Societies. *Current Anthropology, 20*, 798–799.
Das, B., & Mistri, A. (2013). Household Quality of Living in Indian States: Analysis of 2011 Census. *Environment and Urbanization Asia, 4*(1), 151–171.
Kumar, A. (2014). Access to Basic Amenities: Aspects of Caste, Ethnicity and Poverty in Rural and Urban India—1993 to 2008–2009. *Journal of Land and Rural Studies, 2*(1), 127–148.
Kundu, A. (1991). Micro Environment in Urban Planning: Access of Poor to Water Supply and Sanitation. *Economic and Political Weekly, 26*, 2167–2171.
Kundu, A. Bagchi, S., & Kundu D. (1999). Regional Distribution of Infrastructure and Basic Amenities in Urban India. *Economic and Political Weekly, 34*, 1893–1906.
Kundu, A. (2006). Dynamics of Growth and the Process of Degenerated Peripheralization in Delhi: An Analysis of Socio-Economic Segmentation and Differentiation in Micro-Environments. In P. J. Marcotullio and G. McGranahan (Eds.), *Scaling Urban Environmental Challenges: From local to global and back.* Earthscan.
Planning Commission, Government of India. (2008). *Eleventh Five-Year Plan (2007–12), Social Sector* (Vol. II, p. 163). Oxford University Press.
Patel, S. (2013). Upgrade, Rehouse or Resettle? An Assessment of the Indian government's Basic Services for the Urban Poor (BSUP) Programme. *Environment and Urbanization, 25*(1), 177–188.
Sita, K., & Bhagat, R.B. (2007). 'Population Change and Economic Restructuring in Indian Metropolitan Cities: A Study of Mumbai. In Annapurna Shaw (Ed.), *Indian Cities in Transition* (pp. 59–82). Orient Longman.
UN-Habitat. (2006). *Meeting Development Goals in Small Urban Centres: Water and Sanitation in the World's Cities.* Earthscan.

# Chapter 8
# Urban Inequality, Segregation and Urban Crime

## 8.1 Introduction

Historically cities have manifested economic and social inequalities. Plato mentions the presence of rich and poor in the ancient Greek republic of Athens in the fifth century BC, while Engels narrates the appalling living conditions of the working class in nineteenth-century Britain (Hamnett, 2001). In the 1920s, the famous Chicago school attempted to study the urban morphology in Chicago city delineating zones and circles of prosperity and poverty within the city areas. It was an ecological approach to study a city as a relationship between the place and the people evolving with increasing population, economic diversification and multiplication of occupations and professions. The works of the Chicago School show that the forces of competition, invasion and succession play a dominant role in the evolution of the socio-spatial structure of American cities more prominently visible in the emerging contrast between the economically deprived inner city and the prosperous suburbs (Park & Burgess, 1925). An ecological approach applied to the human habitat, known to be the human ecological approach, is premised on the assumption that populations and human institutions are related and exist synergistically within the limits of any natural area or human habitation. Competition for resources among individuals and social groups favours survival of the fittest[1]. The idea of competition and survival of the fittest first appeared in liberal political economy and social philosophy in the writings of Malthus, Adam Smith and Herbert Spencer before entering into Darwin's work on *Origin of Species* (Portugali, 1996). It is a method of natural selection establishing a natural order or a state of equilibrium within which various components act in co-operation and synergy like a living organism. The idea of human ecology has been pervasively ingrained in the mainstream disciplines of sociology and economics rooted in theories of functionalism and economic equilibrium of neo-classical economics, respectively. Functionalists viewed inequalities were an inevitable consequence, and perhaps even desirable for the social order (Davis & Moore, 1945; Fallers, 1973). Similarly, principles of human ecology underscore that

 
R. B. Bhagat and M. I. Hassan, *Urbanisation and Urban Policies in India*, Exploring Urban Change in South Asia, https://doi.org/10.1007/978-981-96-3181-0_8

inequality is natural and the change occurs as a cyclical pattern of transition from social organisation to social disorganisation or economic equilibrium to disequilibrium as a temporal process of transition. In the urban societies, the competition for space and the consequent distribution of population by residence and occupation creates segregation and inequality. Its emphasis on residential segregation is paramount. According to Burgess, differentiation in the residential patterns of economic and social groups gives form and character to the city. Segregation limits development in certain directions, but accentuates certain traits and makes further differentiation (Burgess, 1925). The human ecological approach recognises segregation, functionality and interdependence of various parts in an urban system; however, it does not recognise inequality, power, subjugation and exploitation in the urban social structure constituted by class, caste, race and gender as these are considered to be natural. A city is viewed like a social organism, differentiated within it and composed of various parts as related to the body of the city. On the other hand, any disturbances are viewed as social disorganisation that needs to be ameliorated, but seriously limits our understanding of how inequality is created, sustained or even exacerbated in cities.

Ontologically the natural law of the society based on competition, invasion and succession differs fundamentally from the human law of society based on co-operation, cognition and concern for the future. No doubt that human beings are social organism but differ from its ability to change their habitat based on knowledge, planning, innovations and imaginations. Human beings also created various political, economic and social institutions for their survival, security and progress, and even ensure the survival and progress of the weakest. However, everything is not well with human nature. Human beings are greedy, accumulative and power seeking, dominate fellow human beings and degrade nature. The Sustainable Cities and Communities, i.e. Goal 11 of Sustainable Development Goals (SDG) require a synergy between the humans and the nature perhaps best expressed in the idiom of humanising the nature and naturalising the humans. The city as a form of human habitat needs to be looked from this perspective.

The human ecological approach of Chicago school was criticised by many scholars. Manuel Castells viewed the ideas of Chicago school as naive. According to him, the urban question does not exist outside the capitalist mode of production intimately related to collective consumption such as housing, transport, water and sanitation, health and education, etc., needed for the social reproduction of labour power and means of production. Urbanisation does not have theoretical value in explaining residential segregation or access to amenities or explaining the genesis of inequality, its perpetuation and the consequences, it is simply a spatial form or an empirical reality (Castells, 1976; 1977). On the other hand, David Harvey argued that the process of urbanisation embodies spatial forms and social processes that are intimately related and co-producing each other. Historically industrialisation produced urbanisation, and subsequently urbanisation drove industrialisation. Urbanisation is the product of social surplus, and the city is a place where surpluses are produced and appropriated. Looked from this way urbanisation and cities are class-dominated phenomena reflecting inequalities and segregation in terms of level of income, housing conditions and chances of life (Harvey, 1973). The economic and social processes structured the nature of the urban housing market and, in combination with the existing class and ethnic structure, produced residential patterns. Ghettos, slums, shanties, red light areas, working-class quarters are manifestations of class-divided cities.

Thus, a political economic approach adopts to understand how a city evolves economically as a productive centre, manifests in its consumption spaces and deploys a means of distribution and appropriation very central to the genesis and perpetuation of inequalities, segregation and its various consequences. Thus, cities are not simply a human ecological unit, but economic and political units as well. Cities have been centres of production, engines of economic growth, centres of innovations and wealth creation and seats of civilisation. Who gets, what, and where in the city is very significant to understanding the existing and emerging inequalities closely associated with segregation in the residential pattern in the city. Segregation is a spatial form while inequalities are rooted in the socio-economic process, and manifest in the form of exclusion, violence and crime. This chapter reviews theoretical and empirical works on segregation and inequalities in general and Indian cities in particular. The objective is to understand the relatedness of these two concepts, their magnitudes, manifestations and consequences to throw light on the challenges of inclusion and integration in urban societies.

## 8.2 Inequality and Segregation

What is the ontological basis of inequality? Many believe that it is natural differences like strength, intelligence and genes are the basis of inequalities. However, this is an ecological approach directly or indirectly linked to the Darwinian ideas of survival of the fittest and natural selection. This underlying ontology runs across the dominant social science theories and the epistemology of positivism provides the basis for knowing and understanding the world. On the other hand, many also believe that inequalities are not about natural differences but it is about wealth, income and social status, which in turn determine the corresponding degrees of access to opportunity, skills and knowledge. As the productive capacity of human society increased, so does the amount of wealth and inequalities that have grown historically (Habib, 2015). An economic inequality intersects and recreates social inequalities of varied dimensions related to class, caste, race and gender. An understanding of inequality and its consequences requires a relational methodology to elucidate the dynamics of class, caste, race and gender relationships which are embedded in time and space (Blim, 2023).

One of the fundamental characteristics of urbanisation is that cities create both prosperities and deprivations. While income and wealth increase, so do the crime, pollution and diseases. An interesting study by Bettencourt and West (2010) based on large data on GDP, income, crime and patents of 360 US metropolitan shows that all the good and the bad increases approximately by 15 per cent more than the expected linear growth. The authors call it an interesting city law of 15% rule. This finding reaffirms that in cities the positives and the negatives are two sides of the same social processes and social relationships. However, the lack of data at city level defies such conclusions for India and many other developing countries.

Notwithstanding, it is true that the social and economic inequalities in cities manifest in the form of spatial inequalities. Spatial inequalities do not show simply the disparities among the localities but constitute spatial processes and spatial relationships and determine the use and appropriation of spaces by different social groups and classes. Viewed from this perspective, the spatial and social processes are interrelated and co-produce each other recursively. At city level where we live and work matters as it reflects economic and social status of communities and also determines the access to amenities and resources. Areas within the city are differentiated and segregated along economic class and ethnicity visible in the organisation of neighbourhoods, sectors, enclaves, shanties and slums. Cities are places of extremes showing enclaves of privileges on the on the one hand, pockets of deprivation what researchers call '*enclave urbanism*' seen in different parts of the developing countries (Wissink, 2013). What is emerging is the unequal city, dual city or fractal city? In dual cities, there are two contrasting groups of people live in terms of income, education and skills. There is a large difference in the top and bottom end of occupational hierarchy such as jobs of professional and managerial classes engaged in finance, banking, accounting and IT industry, while others engaged in hotel, restaurant, transport and personal services and many more low paid jobs. Such dualisation in the occupational and social structure of the cities is linked to the process of immigration of low-skilled workers (Castells, 1993; Sassen, 1991). Edward Soja used the term fractal city to denote the polymorphic nature of contemporary cities particularly American cities which no longer could be simply classified in upper class vs middle class or black vs white neighbourhoods, but it is a city of intensified inequalities and social polarisation having a fractured geometrical pattern. In other words, the fractal city is highly differentiated and unequal ethnic and social mosaic of cities (Soja, 2000).

While cities are driving the nation and national economies, they also produce inequalities of various forms and magnitudes. However, inequality in whatever context is ultimately a social problem (Blim, 2023). In fact, inequality limits the human ability of freedom and jeopardises the idea of fraternity and productivity thereby affects the sustainability of cities. A sense of belonging and inclusion is very important for an inclusive and integrated city.

One of the earliest forms of segregation and inequality is seen in the emergence of Jewish Ghettos in the cities of medieval Europe. Forced segregation of Jews spread throughout Europe during the fourteenth and fifteenth centuries. The name *ghetto*, probably derived from an iron foundry in the neighbourhood, was first used in Venice in 1516. The ghettos of Western Europe were permanently abolished in the course of nineteenth century, however, revived during Nazi's Germany during the Second World War. The Warsaw Ghetto was the foremost example during this time[2]. More recently, the term ghettos are being applied to any segregated area of immigrants particularly the African Americans in the American cities signifying the entanglement of race, poverty and inequalities recalling the memories of the ghettos of Europe (Duneier, 2016). The world has also seen the most notorious form of racial segregation known as apartheid legally designed in South Africa which continued until the early 1990s. Although politically it is already abolished, the effects on the socio-economic and spatial structures of cities are reinforced and embedded in racial relations coupled with increased rural–urban divide in the country (Larson, 2019). As such it is evident that the segregation of social groups from one another in space is a manifestation of social and economic inequalities perpetuated through power relationship and economic mechanism of exploitation. The ethnic and racial segregation carries tremendous impacts on health and deprivation, employment prospects, conflict and tensions in the social relationship among the groups involved leading to increased crime and violence. However, the types of consequences that result from segregation depend on group dynamics in the social, political and economic context and a variety of contingent circumstances (Kaplan & Douzet, 2011). In short, the segregation influences the entire economic, political and civic life of the disadvantaged groups who were segregated either through market or designed policy mechanisms. Inequality and segregation in many parts of the world have been rising in the recent years under the neo-liberal regime. A study of European cities by Marcinczak et al., (2016: 358) mentions the following:

> Where state intervention in Europe has long countered segregation, (neo) liberal transformations in welfare states, under the influence of globalisation, have caused an increase in inequality. As a result, the levels of socioeconomic segregation are moving upwards. If this trend were to continue, Europe would be at risk of slipping into the epoch of growing inequalities and segregation where the rich and the poor will live separate lives in separate parts of their cities, which could seriously harm the social stability of our future cities.

The above statement echoes what Martin Luther King, Jr said about segregation about half a century ago.

> Segregation distorts the soul and damages the personality. It gives the segregator a false sense of superiority and the segregated a false sense of inferiority. —Martin Luther King, Jr.[3]

## 8.3 Inequality and Segregation: Indian Perspective

Historically cities have emerged as complex and stratified units, and this is true for India as well. As production increases and surplus accumulates, cities absorb and expand and even new urban centres are created. Technology, division of labour, transport and communication shaped the spatial evolution and nature of urbanisation. The division of labour, its segmentation in the labour market and spatial segregation in residential patterns have imposed enormous challenges to the social mobility of lower classes resulting in the increased economic inequalities contrary to the expectation of the levelling role of urbanisation and economic development. However, it has neither happened in the west nor in the east (Gradín et al., 2021; Piketty, 2014).

In medieval times main towns were wholly or partly enclosed by walls or ditches for defence. Many local rulers also fortified their chief towns. The palace was the focal point around which soldiers, officials and people of higher status including merchants and craftsmen lived, while the least privileged and poor lived outside the city walls. This pattern of urbanism was a pronounced feature in most of the parts of India during the pre-colonial times (Brush, 1962). During the British rule, colonialism shaped the urbanism and urban social structure of India implanting new port cities namely Madras (1644), Bombay (1661) and Calcutta (1690) which are the leading cities of India even today along with the capital city of Delhi. In each of these port cities British created fort area at the centre of the cities designed and occupied by themselves. While they lived mostly inside the fort area, strongly fortified and controlled, the area around it was generally called as 'civil lines' where they built their homes, shops and churches as well as their commercial and administrative headquarters. The armed forces were accommodated nearby in an area called the 'cantonment' or 'camp'. On the other hand, the much larger Indian area that sprang up around the British core was usually referred to as the 'native' or 'black town'. As British control extended across India in the eighteenth and nineteenth centuries, and encompassed many already existing cities, these patterns of spatial segregation by race was found in other cities as well. However, the racial segregation gradually relaxed and more wealthy people and those serving in bureaucracy also started living in the fort and civil line areas (Spodek, 2013). The traces of racial segregation are hardly visible in recent times.

What was called Central Business District (CBD) in Western countries, b*azaar* (market), known as 'chowk'[4] in northern India, served the same purpose during the colonial cities and still do so. The main bazaar is generally located at the centre of the urban centres where richer sections of societies along with trading communities lived there in and around it. The central *bazaars* in large cities are supplemented by neighbourhood bazaars depending upon local demands. Civil lines and cantonments were created in many indigenous towns also where wealth became concentrated. These were planned settlements separated by maintaining a green belt of gardens between them and the indigenous towns. The residential areas of the old urban areas were divided into separate *mohallas* (neighbourhood) based on caste and religious groups. In general, the higher castes and higher status groups lived near the city centre while the lower castes and lower status groups generally lived in the outskirts of urban centres (Bhagat & Sharma, 2001; Grewal, 2014). Historically Indian cities may be characterised in a very simplistic way as evolving with a rich centre and a poor periphery unlike Western cities with a poor centre and a rich periphery. Also, there was no clear-cut separation of residential areas from business or industrial areas in the early decades after independence as generally found in the cities of Europe and America. Brush (1962: 60) observed that 'merchants live the bazaars and small shops; service industries and manufacturing of all kinds are located in predominantly residential areas. In some areas, manufacturing and retail or wholesale trade are carried on in the same establishments'. The separation of living space from the workplace—a characteristic feature of modern industrial society developed under the British (Hasan, 2014: 89).

India got independence from British rule on 15 August 1947 at the cost of partition. There was a huge refugee migration between the newly created borders between Pakistan and India. The gross refugee migration was estimated to be in the range of 12 to 17 million (Hill et al., 2008). Most of the refugees were settled in the bordering provinces of Punjab and West Bengal and also in the National capital Territory of Delhi. The urban morphology and social spaces of many cities of these two provinces and Delhi particularly were hugely influenced. For example, in Calcutta and its outskirts, many of the neighbourhoods around Tollygunge and Jadavpur such as Netaji Nagar, Bijoygarh and Azadgarh were former refugee colonies. In Delhi, areas that are today considered middle-class and upmarket areas such as Malviya Nagar, Kalkaji and Lajpat Nagar were used to settle refugees. Caste and religious identities were marked in the resettlement process, and researchers also argue that partition led to segregation of communities, changes in the demographic composition of communities and ghettoisation of Muslims (Sanyal, 2014).

The partition also led to the transformation of India's urban social space as refugee colonies bourgeoned in many existing cities. New townships known as model towns too began to develop around urban centres in many existing cities of Punjab and Haryana.

The post-independent India has witnessed increasing urbanisation, increased rural and urban divide and the rising regional disparities led by the dominating role of large and megacities. At the city level, existing inequalities and segregations further got accentuated between the high and the low-income areas; posh areas and gated communities on the one hand and slums and squatters on the other. Further, the religious, caste, linguistic and tribal identities also overlapped with the economic structure of the cities reflecting on the spaces of segregations and inequalities.

It is important to mention that the urbanisation in India did not bring about a complete transition of pre-industrial social groups characterised as *gemeinschaft* to *gesellschaft* as noticed in Western Europe. Tonnies (1957) defines *gemeinschaft* as the rural and peasant societies based on personal relationships regulated by traditional social rules and direct face-to-face relations in everyday life living in a contiguous area, while gesellschaft is characterised as a social group based on rationality and self-interest spread over geographically discontinuous areas. The human relations among gesellschaft are impersonal and contractual. Registered societies, trusts, associations, unions and parties are some of the examples of gesellschaft produced by industrial and urban societies. This led to the weakening of the bond of family, ethnicity, kinship and religion paving the way for the emergence of civil society and civic life and the emergence of the idea of citizenship in the modern world. However, the trajectory of such a transition in the social structure of many developing countries including India was different due to colonial urbanisation and the external nature of industrialisation. The expectation that the urbanisation and the concomitant migration would change the norms, practices and relationships within and between the primordial social groups like family, tribe, caste and religion to usher into an urbanism based on the integrative characteristics of civic society inherently premised on secular values. Such expectation continues in social discourse but it is important to realise that the trajectory of social change and urbanism has been different in India due to its historical conditions and social structure.

It may be pertinent to mention that Dr B.R. Ambedkar, Chief Architect of India's Constitution believed that urbanisation and the attendant rural to urban migration will change the caste relations and caste discrimination which has been traditionally rooted in village society. He wrote,

> I hold that these village republics have been the ruination of India...What is the village but a sink of localism, a den of ignorance, narrow-mindedness and communalism? (Moon, 1994: 62)

On the other hand, Mahatma Gandhi viewed urbanisation as a colonial imposition, while celebrating Indian village values. In concurrence with the idea of Ambedkar, Jawaharlal Nehru, the first Prime Minister of India regarded cities playing a significant role in India's modernisation and social transformation (Bonner, 2004). Ironically, the historical evolution of Indian society reveals not only the caste but other primordial social structures also continue to play an important role in the quotidian life in urban India. Historically the institution of caste has been based on the ideology of segregation rooted in the social structure of village communities. However, urbanisation could not erode the spatial segregation of disadvantages of castes evinced in number of studies (Mehta, 1969; Shaban & Aboli, 2021; Thorat et al., 2015; Vithayathil & Singh, 2012). However, it is not to deny that the caste-based social relations in the villages have changed a lot due to urbanisation and increasing the opportunity of migration for salaried jobs and wage labour in cities and towns, the residential spaces continue to exist segregated in cities and towns. Studies also reveal segregations and ghettoisation of religious minorities particularly Muslims in the city (Ghose, 2011; Shaban & Aboli, 2021). The caste and religion-based segregations are embedded in the pronounced economic inequalities between slum and non-slum areas entailing inequities in the access to urban amenities such as water and sanitation, housing, education and health services (Gupta, 2023), as slums consist of a disproportionate share of disadvantaged social groups compared to non-slum areas (Mehra, 2011 and Gupta et al., 2009).

## 8.4 Slum, Segregation and Indian Urban Space

It is paradoxical that urbanisation has been producing both richness and deprivations, and segregation is inherently inbuilt in the urban space. While the wealth of nations rises as they urbanise, so is the inequality and segregation. Segregation in the residential spaces is most glaringly visible in the contrast between slum and non-slum areas. In some cases, slum areas are also surrounded by the presence of gated towers and high-rise buildings in more recent times (Khosla, 2017). Slums are generally looked upon as areas of poverty in popular discourse, but it would be more important to characterise them as areas of exclusions and deprivations. A study based on large-scale quantitative data based on eight cities as a part of National Family Health Survey-3 shows that not all slum dwellers are poor but the poverty is more prevalent in slum areas compared to non-slum areas due to appalling housing conditions, inadequate access to water and sanitation and poor health conditions (Gupta et al.,

2009). Health and housing poverty are more distinct than income poverty in slums (Agarwal, 2011; Goli et al., 2011; Nolan, 2015).

Concepts and operational definitions of slums vary from country to country. UN-Habitat tried to develop a generic definition of slum and also developed an operational measurement of slum households. In an expert group meeting held in Nairobi agreed on the following generic definition:

> A slum is a contiguous settlement where the inhabitants are characterized as having inadequate housing and basic services. A slum is often not recognised and addressed by the public authorities as an integral or equal part of the city. (UN-Habitat, 2003: 6)

Further, UN-Habitat *'defines a slum household in operational terms, as lacking one or more of the following indicators: a durable housing structure; access to clean water; access to improved sanitation; sufficient living space; and secure tenure. The first four rely on conventional definitions; the last is the most difficult to assess and is not currently used in slum measurement'* (UN-Habitat, 2018: 8).

Also, slum is an informal settlement that defies planning, building regulations and municipal approval. In India, some slums are notified or recognised while others are squatters with no security of tenure and constantly face the risk of demolition. UN-Habitat estimates that one-third of urban population lives in slums, while in India as per last census estimate it was about one-fifth of urban population. It is generally accepted that the estimates whether in India or elsewhere are basically lower estimates.

The Census of India defines slum areas as follows:

i. All notified areas in a town or city notified as 'Slum' by State, UT Administration or Local Government under any Act including a 'Slum Act' may be considered as Notified slums.
ii. All areas recognised as 'Slum' by State, UT Administration or Local Government, Housing and Slum Boards, which may have not been formally notified as slum under any act may be considered as Recognised slums.
iii. A compact area of at least 300 populations or about 60–70 households of poorly built congested tenements, in the unhygienic environment usually with inadequate infrastructure and lacking in proper sanitary and drinking water facilities. Such areas may be considered as Identified slums.

In the above definition, slums are compact and overcrowded residential areas not isolated or scattered dwellings unfit for human habitation due to lack of one or more of the basic infrastructures like drinking water, sanitation, electricity, sewerage, streets, etc. (Office of the Registrar General, India, 2019). In 2011 it covered all statutory towns; non-municipal towns/census towns were excluded. In addition to census, the National Sample Survey Office (NSSO) also covers slums which include legally recognised slums and also identified slums of clusters with at least 20 households. Thus, the criteria differ from agency to agency within a country even (Gupta et al., 2009). The nomenclature of slums also varies from one region to another region in India. For example, in Delhi slums are known as 'Jhuggi-Jhopadi', whereas they are called as 'Jhopadpatti or Chawls', in Kanpur 'Ahatas', in Chennai 'Cherries' and in

Bengaluru, slums are known as 'Keris' (Office of the Registrar General, India, 2019). However, the most pathetic deprivation and inequality are neither the slums nor the squatter settlements but a houseless population which live on pavements, under flyovers, railway stations and in open areas. The magnitude of houseless population is hugely underestimated by the census, and hardly there is any other source to reflect on this aspect (Bhagat & Jones, 2014).

The magnitude of the slum varies according to the history and nature of industrial development of the city, land and housing prices, nature of employment, level of wages and public provisioning of housing, etc. Figure 8.1 shows the magnitude of slums as high as 42 per cent in Mumbai and as low as 5 to 8 per cent in Ahmadabad and Bangalore. As urban development is a state subject, the state-specific factors such as politics and perception play important roles in the notification, recognition and identification of slum clusters. Slums are generally located near railway lines, on slopes, in marshy areas, along nalla/drains and river banks, etc. These are environmentally risky and prone to disaster. While the majority of slums are situated on public land such as land belonging to urban local bodies, railways, etc., a size proportion also thrive on private lands (National Buildings Organisation, 2015).

Slums are an integral part of urbanisation and a manifestation of rural to urban migration. However, slums are not only growing due to rural to urban migration only also due to natural increase. Further, there are also intra-urban movement of slum dwellers within the same city and also but migrating between cities and towns.

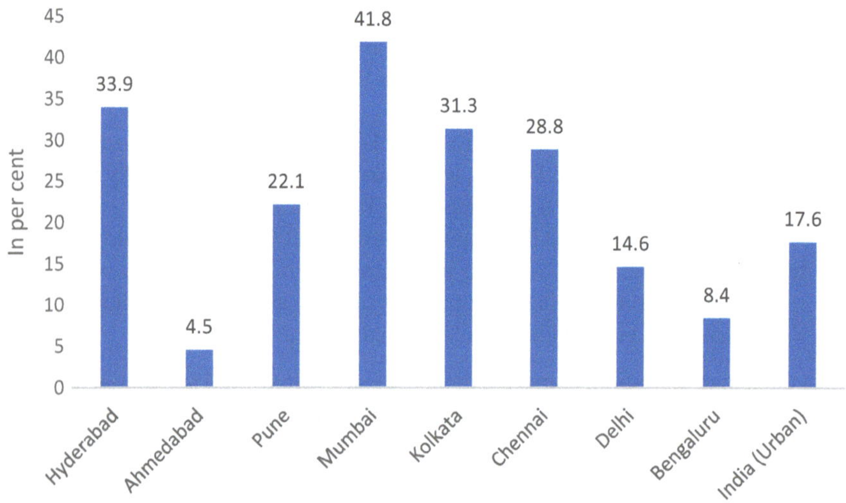


**Fig. 8.1** Percentage of slum population in megacities (Municipal Corporation), India, 2011 (*Source* Census of India, 2011, PCA SLUM: Primary census abstract data for slum, India & States/Uts—Town Level—2011 (https://censusindia.gov.in/nada/index.php/catalog/6190, accessed on 11 November 2024); For population data on cities, PCA: Primary Census Abstract Urban Agglomeration, accessed on 11 November 2024. No complete coverage of slum population in Ahmadabad and Bangalore cities

Migration from non-slum to slum areas is also found within and between the cities and towns (NSO, 2018).

Slums provide cheap housing and serve as a gateway to the large cities through a network of friends and kin who migrated previously and are living there. Cheap housing is also available through rented accommodation in slum areas. The National Family Health Survey-3 shows that one-quarter of slum households in Mumbai and about half of those in Kolkata live in rented accommodation (Gupta et al., 2009:94). The rental market is a thriving business in slum areas of urban India as the data also affirm that a quarter of slum households live in a rented tenement (National Buildings Organisation, 2015).

To conclude, it may be reiterated that slums are the manifestations of divided or dual cities economically produced and socially organised. Urban spaces become complex as economic processes are embedded in caste, race, language and religion. In many cases, the divisions between the social structure of the city and its spatial structure are remarkably corresponding reflecting inequalities, access to basic amenities and struggle of the people.

## 8.5 Inequality, Segregation and Urban Crime

Inequality and segregation are the potential breeding grounds for crime. In classical ecological models, urbanisation has been viewed as leading towards social disorganisation, weakening of social bonds, increasing competition, individualism, alienation and anomie. A generalised version of urbanisation as a sociological phenomenon attributed the personal disorganisation, mental breakdown, suicide, urban crime and corruption to the emerging urbanism in the several sociological writings in the late nineteenth and early twentieth century (Writh, 1938). This also created a perception based on empirical studies that crime rates are higher in urban than rural areas portraying a negative and undesirable nature of urbanisation ignoring the changing nature, magnitude and diversification of crime with increasing urbanisation (Glaeser & Sacerdote, 1999). However, instead of all crimes, if we compare specific crimes occurring in rural and urban areas, such distinction no longer exists (Dhanagare, 1969).

On the other hand, explanations based on the ecological school were mainly sought in the sociological changes and social psychology delinked from the economics of poverty, deprivations, unemployment, low wages and lack of appropriate urban policies and programmes. Further, the ecological school considered migration and demographic characteristics of size and density as defining features of city leading to a popular narrative of migration and demographic change as a root cause of urban miseries and urban crime.

However, the classical ecological school of sociology generally overlooked that urbanisation also creates segregation and segmentation of urban life closely related to increasing wealth and inequalities. A study comparing across the industrialised countries shows that higher inequality is associated with higher crime and lower social trust. The higher crime is the result of the fact that people who experience inequality and wealth surrounding them affect their state of mind, create anxiety and tension and their relationship with others (Courson & Nettle, 2021).

Urban crime is a complex and multidimensional phenomenon. It is not only individualistic and directed against the security of life and property but also structural seen in terms of occurrence of communal, gender-based violence and violence against the weaker sections of society. It is rooted in social relations, social exclusions, marginalisation and the competitive organised social networks and conflicts. Thus, the urbanisation of crime is not only ecological but also greatly influenced by the nature of urban politics and the political economy shaped by economic and social inequalities (Shaban, 2010).

Thus, crime should not be viewed simply as a law enforcement issue but as a wider social issue of social relationship and social existence. Crime-related data and statistics are very important to understand the nature of crime, its intensity and its diversity in order to mitigate the occurrence of crime. In India, the National Crime Record Bureau (NCRB) under the Ministry of Home Affairs, Govt. of India compiles crime-related data right from the police station level (https://ncrb.gov.in). It also provides crime statistics for all metropolitan cities with population of more than 1 million. As the nature of crime has been changing with increasing urbanisation and economic development, crime statistics need to be evolving in its typology and classifications. In this respect crime against women, children and crime against senior citizen assumes special significance for the safe city. Crime statistics has also been provided for cyber crime and environment-related crime along with physical and economic offences (NCRB, 2022). Although data has several problems, urban researchers need to look into them carefully to understand the dynamics of city and urban space. It also needs to be examined and supported by other sources of data and information as well.

Safe city is a precondition for the inclusive and sustainable cities. This has been emphasised in Sustainable Development Goal (SDG) 11 on Sustainable Cities and Communities recognising the need for the capacity building and institutional strengthening for inclusive, participatory and sustainable urbanisation and urban planning. Further, SDG Goal 16 explicitly recognises reducing inequality and promoting inclusion as a key strategy for reducing crime and violence and promoting safe cities. Such strategies are crucially linked to enhance resilience at the individual and community level against the external forces of globalisation and to strengthen engagement of communities in the enforcement of rule of law (UNODC, 2016). Protection of human rights and community engagement are vital along with socio-economic policies of development and inequality reduction for the safe, inclusive and prosperous cities premised on the principles of sustainable urbanisation guided by the ethical concern in the SDG slogan of 'Leave No One Behind'.

## 8.6 Concluding Remark

It is very much evident that the process of urbanisation has been creating wealth through the production and commodification of space, but also creating inequality, segregation and exclusions leading to insecurities and unrest undermining the basic foundations of inclusive, safe and sustainable cities. Consequently, it raises a voice for the *right to the city* or right to the space. Cities cannot be left to grow organically but need to be consciously built, owned and occupied. It cannot happen without local democracy, participatory planning and collective ownership. Right to the city is an urban imagination of widening freedom and capabilities, promoting a sense of dignity through equality to everyone, and fraternity based on urban inclusion. It is about the nation building by fulfilling the aspirations of urban citizenship. In the words of a visionary urban thinker, David Harvey reasons that 'the *Right to the City* is far more than the individual liberty to access urban resources; it is a right to change ourselves by changing the city. The lack of freedom to make and remake our cities and ourselves is one of the most precious yet most neglected of our human rights' (Harvey, 2008).

**Notes** 1. **Charles Darwin used the term, *survival of the fittest***, in the fifth edition of *On the Origin of Species* published in 1869. Darwin borrowed the term from English sociologist and philosopher Herbert Spencer, who first used it in his 1864 book *Principles of Biology*. https://www.britannica.com/science/survival-of-the-fittest, accessed on 5 March 2024.

2. Britannica, T. Editors of Encyclopaedia (Invalid Date). *Ghetto. Encyclopaedia Britannica*. https://www.britannica.com/topic/ghetto, Accessed on 16 March 2024.

3. *Martin Luther King Jr, Letter from Birmingham Jail; April 16, 1963*, African Studies centre, University of Pennsylvania, https://www.africa.upenn.edu/Articles_Gen/Letter_Birmingham.html, accessed on March 25, 2024.

4. A chowk is a square though in many cases it would be a rectangle. It is a central market-place with many streets joining it. The most important is Chandni Chowk in Old Delhi (erstwhile Shahjahanabad) (see Hasan, 2014).

## References

Agarwal, S. (2011). The State of Urban Health in India; Comparing the Poorest Quartile to the Rest of the Urban Population in Selected States and Cities. *Environment and Urbanization, 23*, 13–28.

Bettencourt, L., & West, G. (2010). A Unified Theory of Urban Living. *Nature, 467*(21), 912–913.

Bhagat, R. B., & Sharma, S. (2001). Quality of Life, Gender and Fertility Behaviour in Urban Haryana: A Case Study of Rohtak City. *Annals of the National Association of Geographers, India, 21*(2), 47–57.

Bhagat, R. B., & Jones, Gavin, W. (2014). Demographic Dynamics of Mega-Urban Regions: The Case of Mumbai. *Demography India, 43*(1&2), 71–94.

Blim, M. (2023). *Inequality (Oxford in Handbook of Cultural Anthropology Oxford).* https://doi.org/10.1093/acrefore/9780190854584.013.398

Bonner, P. L. (2004). 'Migration, Urbanization and Urban Social Movements in Twentieth Century in India and South Africa. *Studies in History, 20*(2), 215–236.

Burgess, E. W. (1925). The Growth in the City: An Introduction to a Research Project. In R. E. Park, & E.W. Burgess (Eds.), *The City*. University of Chicago Press.

Brush, John, E. (1962). The Morphology of Indian Cities. In Roy Turner (Ed.), *India's Urban Future* (pp. 57–70). University of California Press.

Census of India. (2011). PCA SLUM: Primary Census Abstract Data for Slum, India & States/ UTs—Town Level—2011 PC11_PCA-SLUM Primary Census, https://censusindia.gov.in/census.website/data/census-tables. Accessed on March 31, 2024.

Castells, M. (1977). *The Urban Question: A Marxist Approach.* Edward Arnold, French original published.

Castells, M. (1976). Theory and Ideology in Urban Sociology. In C.G. Pickvance (Ed.), *Urban Sociology: Critical Essays.* Methuen & Company Ltd and Transtock Publications, (First published).

Castells, M. (1993). European Cities, the Informational Society, and the Global Economy. *Tijdschrift Voor Economische En Sociale Geografie, 84*, 247–257.

Courson, B. D. & Nettle, D. (2021). Why Do Inequality and Deprivation Produce High Crime and Low Trust? *Scientific Reports, 11*, Article number 1937, https://www.nature.com/articles/s41598-020-80897-8

Davis, K., & W. Moore. (1945, April). Some Principles of Stratification. *American Sociological Review, 10*(2), 242–249.

Dhanagare, D. N. (1969). Urbanism and Crime. *Economic and Political Weekly, 4*(28/30), 1239–1242.

Duneier, M. (2016). *Ghetto: The Invention of a Place, the History of an Idea.* Farrar.

Fallers, L. (1973). *Inequality: Social Stratification Reconsidered.* University of Chicago Press.

Ghose, J. (2011). Muslim Neighbourhoods and Segregation in the city. In M. Zerah, V. Dupont, S. T. Lama-Rewal, & M. Faetanini (Eds.), *Urban Policies and the Right to the City in India: Rights* (pp. 43–47). Delhi.

Glaeser, E. L., & Sacerdote, B. (1999). Why Is there More Crime in Cities? *Journal of Political Economy, 107*(6), S225–S258.

Gradín, C., Leibbrandt, M., & Tarp, F. (2021). *Inequality in the Developing World.* WIDER Studies in Development Economics Oxford University Press.

Grewal, R. (2014). Urban Morphology Under Colonial Rule. In I. Banga (Ed.), *The City in Indian History: Urban Demography* (pp. 173–190). Society and Politics, Manohar.

Gupta, D. (2023). Urban Space, Residential Segregation and Access to Amenities: A study on Kolkata, PhD Thesis, International Institute for Population Sciences, Mumbai.

Gupta, K., Arnold, F., & Lhungdim, H. (2009). *Health and Living Conditions in Eight Indian Cities*, NFHS-3, India, 2005–06, International Institute for Population Sciences, Mumbai.

Goli, S., Arokiasamy, P., & Chattopadhayay, A. (2011). Living and Health Conditions of Selected Cities in India: Setting Priorities for the National Urban Health Mission. *Cities, 28*, 461–469.

Habib, I. (2015). Inequalities: A Social History. *Social Scientist, 43*(1/2), 3–8.

Harvey, D. (2008). The Right to the City. *New Left Review,* 53 (September–October), 23–40.

Hill, K., Selzer, W., Leaning, J., Malik, S. J., & Russell, S. S. (2008). The Demographic Impact of Partition in the Punjab in 1947. *Population Studies, 62*(2), 155–170.

Hamnett, C. (2001). Social Segregation and Social Polarisation. In R. Paddison (Ed.), *Handbook of Urban Studies* (pp. 162–176). Sage Publications.

Hasan, N. S. (2014). The Morphology of a Medieval Indian City: A Case Study of Shahjahanabad. In Indu Banga (Ed.), *The City in Indian History: Urban Demography, Society and Politics* (pp. 87–97). Manohar.
Harvey, D. (1973). *Social Justice and the City*. Blackwell Publishers.
Tonnies, F. (1957). *Community and Society: Gemeinschaft und Gesellschaft*. (F. Toennies, Trans.). In Charles P. Loomis (Ed.). Michigan State University Press.
Kaplan, D. H., & Douzet, F. (2011). Research in Ethnic Segregation III: Segregation Outcomes. *Urban Geography, 32*(4), 589–605. https://doi.org/10.2747/0272-3638.32.4.589
Khosla, R. (2017). India's Urban Landscape: Black Towns of the 21st Century. *Economic and Political Weekly, 52*(1), 92–101.
Larson, Z. (2019, August). South Africa: Twenty-Five Years Since Apartheid. *Origins: Current Events in Historical Perspective*. https://origins.osu.edu/article/south-africa-mandela-apartheid-ramaphosa-zuma-corruption?language_content_entity=en
Marcinczak, S., Musterd, S., Ham, M. V., & Tammaru, T. (2016). Inequality and Rising Levels of Socio-Economic Segregation: Lessons from a Pan-European Comparative Study. In T. Tammaru et al. (Eds.), *Socio-Economic Segregation in European Capital Cities: East meets West*. Routledge, New York.
Mehra, D. (2011). Caste and Class in Indian Cities: Habitation, Inequality and Segregation. In M. Zerah, V. Dupont, S. T. Lama-Rewal, & M. Faetanini (Eds.), *Urban Policies and the Right to the City in India: Rights* (pp. 39–42). Delhi.
Mehta, S. (1969). Patterns of Residence in Poona, India, by Caste and Religion: 1822–1965. *Demography, 6*(4), 473–491.
Moon, V. (1994). Draft Constitution—Discussion. In *Dr Babasaheb Ambedkar Writings and Speeches*. Government of Maharashtra.
National Buildings Organisation. (2015). *Slums in India: A Statistical Compendium*. Ministry of Housing and Urban Poverty Alleviation, Government of India.
NCRB (National Crime Records Bureau). (2022). *Crime in India 2022, Statistics* (Vol. I). Ministry of Home Affairs, Govt. of India. https://ncrb.gov.in.
Nolan, L. B. (2015). Slum Definitions in Urban India: Implications for the Measurement of Health Inequalities. *Population and Development Review, 41*(1), 59–84. https://doi.org/10.1111/j.1728-4457.2015.00026.x
NSO (National Statistical Office). (2018). *Drinking Water, Sanitation, Hygiene and Housing Condition*, NSS 76th Round, Ministry of Statistics and Programme Implementation, Govt. of India.
Office of the Registrar General, India. (2019, December 4). Census of India 2021, Circular No. 7, Formation and Identification of Slum Enumeration Blocks for Slum Demography, No 9/46/2019, CD (CEN), Office of the Registrar General, India.
Park, R. E., & Burgess, E. W. (1925). *The City*. University of Chicago Press.
Piketty, T. (2014). *Capital in The Twenty-First Century*. The Belknap Press of Harvard University Press.
Portugali, J. (1996). Notions Concerning the Nature of World Urbanization. *Progress in Planning, 46*(3), 145–162. https://doi.org/10.1016/0305-9006(96)88867-2
Soja, E. (2000). *Postmetropolis: Critical Studies of Cities and Regions*. Blackwell.
Sanyal, R. (2014). *The Continuing Legacy of Partition in India's Urban Spaces*, London School of Economics. https://blogs.lse.ac.uk/southasia/2014/11/17/. Accessed on March 24, 2024.
Sassen, S. (1991). *The Global City*. Princeton University Press.
Shaban, A., & Zinat, A. (2021). Socio-spatial Segregation and Exclusion in Mumbai. In M. van Ham et al. (Eds.), *Urban Socio-Economic Segregation and Income Inequality*. The Urban Book Series,https://doi.org/10.1007/978-3-030-64569-4_8
Shaban, A. (2010). *Mumbai: Political Economy of Space and Urban Crime*. Orient-Blackswan.
Spodek, H. (2013). City Planning in India under British Rule. *Economic and Political Weekly, 48*(4), 53–61.

Thorat, S., et al. (2015). Urban Rental Housing Market: Caste & Religion Matters in Access. *Econ Polit Wkly, 50*(26 & 27), 47–53.

UN-Habitat. (2003). *Guide to Monitoring Target 11: Improving the Lives of 100 million Slum Dwellers*. United Nations Human Settlement Programme (UN-Habitat), Nairobi.

UN-Habitat. (2018). *SDG Indicator 11.1.1 Training Module: Adequate Housing and Slum Upgrading*. United Nations Human Settlement Programme (UN-Habitat).

UNODC (United Nations Office on Drugs and Crime). (2016). Governing Safer Cities: Strategies for a Globalised World. https://www.unodc.org/documents/justice-and-prison-reform/SaferCities.pdf. Accessed on April 5, 2024.

Vithayathil, T., & Singh, G. (2012). Spaces of Discrimination: Residential Segregation in Indian Cities. *Economic and Political Weekly, 47*, 60–66.

Wirth, L. (1938). Urbanism as a Way of Life. *American Journal of Sociology, 14*, 1–24.

Wissink, B. (2013). Enclave Urbanism in Mumbai: An Actor-Network-Theory Analysis of Urban (dis) Connection. *Geoforum, 47*, 1–11.

# Chapter 9
# Urban Policies and Programmes

## 9.1 Introduction

Sustainable and inclusive urbanisation is a critical factor in order to achieve SDG goals. Goal 11 is specifically concerned with Sustainable Cities and Communities. In this respect, urban policies and programmes are very important to achieve this goal. This chapter will make a distinction between urban development policies which are focused on cities and their problems, and urbanisation policy which addresses the economic, social and environmental links between urban and rural areas by strengthening national and regional development planning. The purpose of urbanisation policy is to reduce the rural and urban divide and act as a vehicle for rural development. Urban development policies and programmes have distinct bias towards towns and cities. Although, there are some programmes launched under Rurban Mission that fall under the urbanisation policy, its architecture and funding have very limited impact on the life of people in the countryside.

A vision of national planning as an instrument of independent India's development began in sessions of the Indian National Congress in the 1930s. The Indian National Congress formed the National Planning Committee under the chairmanship of Jawaharlal Nehru in 1938, although the work of committee could not be completed due to intensified freedom struggle in subsequent years. Against this backdrop, the 'Bombay Plan' a blueprint of India's economic development was proposed by a group of industrialists and technocrats in 1944 (Sanyal, 2018). However, the actual planning process started with the establishment of Planning Commission in March 1950 (Planning Commission, 2010). On the other hand, when we look at India's history of development, we find that the urban concerns were precursor to the national development strategies which started by Lord Mayo's resolution in 1870 on local government paving the way for municipal governance in India. The purpose was to give limited power to Indian leaders to contain the national upsurge against the British Raj. The emphasis was on the maintenance of essential services like sanitation and water supply in cities and towns, and not on urban development. Many



R. B. Bhagat and M. I. Hassan, *Urbanisation and Urban Policies in India*, Exploring Urban Change in South Asia, https://doi.org/10.1007/978-981-96-3181-0_9

great national leaders like Pherozeshah Mehta, Surendranath Banerjee, Lajpat Rai, G.K. Gokhale, Vallabhbhai Patel, C. R. Das, Jawaharlal Nehru and Subhas Chandra Bose were associated with municipal governance (Bose, 1980).

A fundamental shift towards urbanisation is visible in the last few decades with emerging dominance of urban centres in general and metro cities in particular. India has launched a policy of economic liberalisation since 1991. During the post-liberalisation phase, the importance of cities and urban centres has been growing in India's economic development, and it is also being emphasised that 9 to 10 per cent of growth in GDP depends fundamentally on making Indian cities more livable and inclusive (Planning Commission, Govt. of India, 2008: 394). The Planning Commission documents of the Twelfth Five-Year Plan (2012–2017) stressed this shift which continued to look at the positive role of cities and the process of urbanisation. This shift is not only economic but decisively demographic and political as well. Not only a huge population lived in urban areas but also conscious, educated and young urban voters now play an important role as both parliamentary and legislative constituencies are getting more urbanised (Times of India, 2008: 16).

An appropriate urban development strategy along with efforts to minimise the differences between town and countryside and by providing infrastructural facilities connecting the far-flung areas that the development could be more inclusive and fulfil the aspirations of the people. It is thus very important to emphasise that urbanisation and urban policies have a critical role in shaping the development outcomes (Bose, 1980).

## 9.2 Urban Policy and Programmes in Five-Year Plans

In India, the cities and towns have grown on their own historically. Policy and programmes to some extent affected the big urban centres, but the small cities and towns and non-metroploitan areas have largely been remained unaffected by urban policy and programmes (Shaw, 1996: 224). On the other hand, although urban development is a state subject, the policy and programmes are formulated by the central government. The central government's five-year plan reflects much of their concerns and also the policy and programmes followed by both the central and state governments through various initiatives and funding status stated in the plan documents. Therefore, five-year plan documents were perhaps the best sources to analyse the urban policy and programmes until the Planning Commision was abolished in 2014.

About 7.5 million refugees crossed to the Indian side in the event of partition of the country in 1947 (http://planningcommission.nic.in/plans/planrel/fiveyr/1st/1planch38.html).

In the first (1951–1956) five-year housing was a major concern in urban areas and also resettlement of the refugees in the rural areas. In both the first and second (1956–1961) five-year plans urbanisation was, by and large, not a significant issue, but concerns were raised about haphazard growth of urban areas and the regulation of land use and control of land prices. It also advised for the enactment of town and country planning legislation in all states to enable the drawing up of Master Plans for the closer regulation of urban land (Ramachandran, 1989: 570). Also, during the first two five-year plan periods, various institutions and organisations were created and set up. For example, the central government set up the Town and Country Planning Organisation, the National Building Organisation and Delhi Development Authority during this period. Effort has been made to prepare the Delhi Master Plan which served as a model for city planning in other states as well (Shaw, 1996: 225).

The Third Five-Year Plan (1961–1966) was a turning point in India's history of urban development and planning. It recognised the importance of cities and towns in balanced regional development and advised that urban planning adopt a regional approach. It emphasised the need for urban land regulation, checking of urban land prices and also preparation of Master Plans for the big cities. It made clear that the preparation of Master Plan is the responsibility of state and local governments (http://planningcommission.gov.in/plans/planrel/fiveyr/index6.html). The majority of the states introduced town planning legislation modelled on the pattern of British town planning designs and practices during the Third Five-Year plan (Shaw, 1996: 226). The Fourth Five-Year plan (1969–1974) continued to emphasise the regional and urban development initiatives in the Third Five-Year Plan, and development plans for many urban centres were undertaken including regional studies and planning of metropolitan regions around Delhi, Greater Bombay and Calcutta were initiated. http://planningcommission.nic.in/plans/planrel/fiveyr/4th/4planch19.html). The development of new state capitals like Chandigarh, Gandhinagar, Bhopal and Bhubaneswar was speeded up through special grants by the central government. In a very significant development during this plan period, the Government of Maharashtra passed the Mumbai Metropolitan Development Act in 1974 and the Mumbai Metropolitan Region Development Authority (MMRDA) was established in 1975. Also, during this period, the Housing and Urban Development Corporation (HUDCO) was established to fund projects of the urban local bodies, housing boards and other organisations. One of the special features of the Fourth Plan was that metropolitan cities and cities of national importance got special financial commitment which continued in the Fifth Five-Year Plan (1974–79) as well (http://planningcommission.nic.in/plans/planrel/fiveyr/welcome.html). There was an apparent favour to the big cities until Fifth Five-Year Plan, although rhetoric of urban and industrial decentralisation was repeated plan after plan (Shaw, 1996: 227).

The Sixth Five-Year Plan (1978–1983) stressed the need to develop small and medium-size towns (less than 100 thousand). A scheme of Integrated Development of Small and Medium Towns (IDSMT) was launched in 1979 by the central government which aimed to improve the economic and physical infrastructure and to provide essential facilities and services in about 200 small and medium towns. (http://planningcommission.gov.in/plans/planrel/fiveyr/index6.html).

It was envisaged that an equal amount will also be provided by the respective state governments. However, it may be noted that according to the 1981 census there were 3029 small and medium towns (population less than 100 thousand) out of 3245 urban centres/urban agglomerations (Bhagat & Mohanty, 2009). Environmental improvement of slums was another thrust area of the sixth plan. It recognised the fact that a particular area becomes a slum more because of poor environmental conditions, poor drainage and sanitation, rather than the poor state of structures. It also suggested to accord priority to restrict the growth of population in the large cities and reemphasised the need of granting incentives to shift industries to new urban centres and also to put a ban on the setting up of new industries within certain limits of large cities as announced in the Industrial Policy Statement. In some large cities, it also suggested to move central and state government offices, banks and other institutions to the less congested areas of the city. The Seventh Five-Year Plan recognised that municipal governments are politically, administratively and financially weak and there is a need to democratise and devolve power to them. It emphasised the need of municipal reforms for their financial viability and efficient administration. It was visualised that there was a need of more private investment to boost industrial activities in and near the small and medium towns so that migration to big cities for employment purposes could be retrained. The ISDMT scheme continued in the Seventh Plan (1985–1990) as well but the fund allocated was 88 crores which is lower than the amount earmarked for the Sixth Plan period. Also, the eligibility for being small and medium town was raised from less than one hundred thousand population to less than 3 hundred thousand populations. Supply of land was thought to be very crucial for urban development and City Development Authorities were advised to meet the need of land supply, particularly, the requirement of serviced land and check the speculative land market. The Seventh Plan also continued its emphasis on Environmental Improvement of Urban Slums (EIUS) and advised to ensure land security to the slum dwellers so that they may improve their housing conditions accessing the market resources (http://planningcommission.gov.in/plans/planrel/fiveyr/index7.html). A Centrally Sponsored Scheme known as Urban Basic Services was introduced in 1986 for the welfare of low-income families with the help of UNICEF. It aimed promoting child survival and development. The services, meant to be delivered, included environmental sanitation, primary health care, pre-school learning, vocational training and convergence of other social services at the slum level (http://planningcommission.nic.in/plans/planrel/fiveyr/8th/vol2/8v2ch13.htm).

However, this programme was merged with EIUS and known as Urban Basic Services for the Urban Poor (UBSP) since 1990–1991. A new programme known as the Nehru Rojgar Yojana was launched in October 1989 with an aim to reduce urban poverty. It is important to highlight that urban policy and programmes until the Sixth Plan mainly addressed the problems like urban housing, slum improvement and upgradation, water supply and sanitation, preparation of Master Plans, development of small and medium towns, strengthening of urban legislation and municipal administration, etc. The Seventh Plan made a new beginning by explicitly recognising the problems of urban poor which were seen to be linked with the creation of employment opportunities (http://planningcommission.nic.in/plans/planrel/fiveyr/8th/vol2/8v2ch13.htm).

There were some important institutional developments during the Seventh Plan period which shaped the urban development policy and planning in India. The 74th Constitution Amendment Bill 1992 was passed by both houses of Parliament and came into force in March 1993. The country saw a serious economic crisis at the end of the Seventh Plan period, the annual growth of GDP plunged to about 3 per cent during 1990–1992 from about 6 per cent during the Seventh Plan period. There was a catastrophic balance of payment crisis and central covernment initiated a new economic policy of delicensing, deregulation, disinvestment of the public sector and the liberalisation process in 1991. The years 1990–1991 and 1991–1992 were not covered under any five-year plans as Eighth Plan started off in 1992–1993 covering the period 1992–1997. The Eighth Plan was practically an extension of the Seventh Plan as most of the urban development programmes were retained. Both the Seventh and Eighth Plans made a departure from earlier plans by resorting to employment generation and concerns for the urban poor. The Nehru Rozgar Yojana (NRY) was meant for households below the poverty line and SCs and STs continued in Eighth Plan. It has three components namely (i) The Scheme of Urban-Micro Enterprises (SUME) designed to encourage unemployed and under-employed youth to take up self-employment ventures in all urban settlements. This was a loan-cum-subsidy scheme with components for training for the beneficiaries and infrastructure development like common facility centres, job centres, design centres, technology upgradation and marketing, etc., (ii) The Scheme of Urban wage Employment (SUVVE) was designed to provide employment to the urban poor through the creation of socially and economically useful assets in the low-income neighbourhoods in towns with a population below one hundred thousand. The creation of assets included in this scheme were: community centres, common worksheds, common selling places for the poor, paving of lanes, low-cost water supply, construction of drains and sewers, pay and use community baths cum latrines and ponds, etc., (iii) The Scheme of Housing and Shelter Upgradation (SHSHU) aimed at providing employment for persons involved in housing and building activities such as masons, sanitary and other construction workers. The scheme is operative in urban settlements having a population between one hundred thousand and 2 millions with a relaxation in the population criteria for hilly states, UTs and new industrial townships. HUDCO is monitoring this component of NRY and also provides the requisite institutional finance.

A programme known as Prime Minister's Integrated Urban Poverty Eradication Programme (PMIUPEP) was initiated in November 1995 during the Eighth Plan especially for the small towns where the poverty was graver due to lack of resources. This scheme was approved for the period up to the year 2000 and applied to urban centres with a population ranging between 50 thousand and one hundred thousand subject to the condition that elections to local bodies have been held.

During the Eighth Plan also the Mega City Scheme was introduced in 1993–1994 covering five megacities of Mumbai, Calcutta, Chennai, Bangalore and Hyderabad. Another centrally sponsored Accelerated Urban Water Supply Programme (AUWSP) was launched in 1993–1994 during this plan. It aimed at providing water supply in towns with a population of less than 20,000. Also, the IDSMT scheme was revamped to dovetail its activities of infrastructure development programmes for boosting employment generation for diverting migration from the big cities to the small and medium towns. The role of the small and medium towns was envisaged as developing growth centres along with Master Plan of big cities and regional plan a recognising the importance of urbanisation as a development strategy. (http://planningcommission.nic.in/plans/planrel/fiveyr/8th/vol2/8v2ch13.htm).

The Ninth Five-Year Plan (1997–2002) emphasised that state-specific urbanisation strategy should be prepared for establishing synergy among various urban development programmes. It is argued that national programmes were top down and often lacked feedback from the states. At times, programmes were ad-hoc and names changed frequently. Further, there was also a multiplicity of programmes directed to the same target group and the coordination among the implementing organisations was lacking. Thus, it was felt that a new policy framework should be in consonance with the provisions of the 74th Constitution Amendment to provide an opportunity to the urban local bodies during the Ninth Plan. Although most of the programmes undertaken in the Eighth Plan continued in the Ninth Plan, the emphasis was placed more on decentralisation and financial autonomy of the urban local bodies with an aim to promote competitiveness and efficiency through market-based interventions. Several earlier programmes were merged to form a new programme called the Swarna Jayanti Shahari Rozgar Yojana (SJSRY) with effect from 1 December 1997. This programme has two sub-schemes, namely, (i) Urban Self-Employment Programme and (ii) Urban Wage Employment Programme. The programme, however, has been discontinued from the financial year 2005–2006. It is important to mention that the funding for small and medium towns under IDSMT continued to be low (http://planningcommission.nic.in/plans/planrel/fiveyr/9th/vol2/v2c3-7.htm). Under a low level of financial commitments, it was not possible for these towns to act as growth centres, spread urbanisation and promote regional development in the country.

The Tenth Five-Year Plan (2002–2007) recognised the fact that urbanisation played a key role in accelerating economic growth in the 1980s and 1990s as a result of economic liberalisation. It also noted that programmes such as the Mega City project for five selected cities, the Integrated Development of Small and Medium Towns (IDSMT), and the Accelerated Urban Water Supply Programme (AUWSP) have shown limited success. It also noted that the coverage and the amount of central assistance in the past have been uneven and inadequate, both because of procedural issues as well as limited budgetary allocations. The existing schemes for assistance for infrastructure such as the IDSMT and the Mega City Project leave a significant number of cities between them without any central support. Also, the scheme like IDSMT could not succeed due to a number of reasons namely (i) lack of capacity for implementation, (ii) non-availability of matching state's share and (iii) non-availability of unencumbered land for the projects (Planning Commission, 2008: 397).

It was also accepted the fact that parastatal agencies and bodies such as development authorities need to play a supportive role and even partner with Urban Local Bodies (ULBs) rather than taking over functions which properly belong to the ULBs. Overall, the Tenth Five-Year Plan stressed that without strengthening the democratic structure and institutional building of the ULBs, the goal of urban development could not be achieved (http://planningcommission.nic.in/plans/planrel/fiveyr/10th/volume2/v2_ch6_1.pdf).

In order to revitalise the urban development strategies, the central government launched a major initiative named as Jawaharlal Nehru National Urban Renewal Mission (JNNURM), in December 2005, to give focused attention to the integrated development of urban infrastructure and services initially in select 63 mission cities with emphasis on provision of basic services to the urban poor, including housing, water supply, sanitation, road network, urban transport, the development of inner/old city areas, etc. JNNURM was considered as an urban renewal programme which aims to encourage reforms and planned development of the identified cities of national importance. Various programmes like Mega City, IDSMT, etc., were merged with JNNURM, and IDSMT was renamed as UIDSSMT (Urban Infrastructure Development Scheme for Small and Medium Towns) under it.

Under JNNURM it was made mandatory for each city to formulate its City Development Plan (CDP) and brings out a long-term vision for the city and its course of development. It also put several conditionalities of urban reforms like repealment of Land Ceiling Act 1976, empowerment of ULBs, capacity building and improvement in municipal accounting. It also aimed to leverage and incorporate private sector in the development and financing of projects through Public Private Partnership (PPP) The employment generation and poverty alleviation under SJSRY were reemphasised in Eleventh Plan with major emphasis on skill development, self-help groups and micro finance for self-employment. The approach of the 'Kudumbshree' model of Kerala and other best practices in different parts of the country were advised to be adopted for suitable application under SJSRY (Planning Commission, 2008).

Further, during the Tenth Plan a scheme known as Provision of Urban Amenities in Rural Areas (PURA) was launched in 2004–05, by the Ministry of Rural Development to encourage the development of identified growth centres for the working population in the rural areas and prevent their migration to the urban areas. To begin with the development of clusters on a pilot basis has been envisaged at seven locations in the country (Planning Commission, 2008: 158). The budget allocated was just 40 to 50 million rupees per cluster. The pilot phase, however, did not produce desired results because of lack of business plan, limited attention to economic activities, lack of ownership and institutional structure and lack of convergence of various rural development programmes. In the Eleventh Five-Year Plan (2007–2012) a restructured PURA was launched as a demand-driven programme through Public–Private Partnership (PPP) between Gram Panchayats and private sector partners with support from central and state governments (Planning Commission, 2013).

The Eleventh Five-Year Plan (2007–2012) also introduced some of the innovative changes in urban policy and programmes. The key urban development strategies were as follows:

i. Strengthening urban local bodies through capacity building and better financial management.
ii. Increasing the efficiency and productivity of cities by deregulation and development of land.
iii. Dismantling public sector monopoly over urban infrastructure and creating a conducive atmosphere for the private sector to invest.

iv. Establishing autonomous regulatory framework to oversee the functioning of the public and private sector.
v. Reducing the incidence of poverty.
vi. Using technology and innovation in a big way.

The midterm evaluation of the Eleventh Five-Year Plan admitted that urban development and urban renewal require huge funds under JNNURM. (http://planningcommission.nic.in/plans/mta/11th_mta/chapterwise/chap18_urban.pdf).

It was observed that there were several barriers to the implementation of JNNURM notably among them is the failure to mainstream urban planning, incomplete reforms and slow progress in project implementation. Delays in securing land for projects and obtaining approval from various regulatory authorities were other reasons for its slow progress. It was found that in many cities urban planning is yet to be started and needs to be strengthened and made participatory in many. Invariably peri-urban areas outside the city limits are left by the very nature of planning narrowing down to a city development plan (Bhagat, 2014).

The Rajiva Awas Yojana (RAY) was launched on 2 June 2011 with the vision of creating a slum-free India as a pilot project for a period of two years. The central government finally approved the RAY on 3 September 2013 for the period 2013–2022. This scheme specifically aimed to support state and city governments to upgrade slums and assign titles to their residents, and to plan to accommodate the envisaged growth as India's rapid urbanisation continues, so that more slums are not formed. The scheme was applicable to all slums within a city, whether notified or non-notified (including identified and recognised), whether on lands belonging to the Central Government or its Undertakings, Autonomous bodies created under the Act of Parliament, State Government or its Undertakings, Urban Local Bodies or any other public agency and private sector. It is also applicable to 'urbanised villages' inside the planning area of the city, urban homeless and pavement dwellers (http://mhupa.gov.in/ray/ray_index.htm). The urban poverty reduction and employment generation programme under SJSRY is proposed to be revamped under the National Urban Livelihood Mission during the Twelfth Five-Year Plan.

As the Twelfth Five-Year Plan was truncated due to the Planning Commission abolished in 2014 and NITI Aayog was formed in its place, the most of urban development programmes have been recast in new avatars since then.

## 9.3 Urban Development Strategies—Smart Cities Mission and Atal Mission for Rejuvenation and Urban Transformation (AMRUT)

The paradigm of smart cities appeared in the late 80s as a means to visualise urban context, and since then they evolved fast in different contexts (Anthopoulos & Vakali, 2012). According to Townsend (2014), 'smart cities are places where information technology is welded to address problems old and new. The old city of concrete, glass and steel now conceals a vast underworld of computer and softwares. The new city, on the other hand, is a digital upgrade to our built legacy giving rise to a new wind of city'—we may call it a smart city. There is also a perspective that tries to look at the smart city, not as a whole but as part of it. For example, it is argued that whatever smart city means but not all spaces of the city will be equally smart. It means that some places, people and activities will have privileges over others (Shelton et al., 2015). A global IT company IBM thinks that 'in the twenty-first century, cities compete globally to attract both citizens and businesses. A city's attractiveness is directly related to its ability to offer the basic services that support growth opportunities, build economic value and create competitive differentiation. Potential inhabitants, of both the commercial and residential variety, are a discriminating lot, and they are looking for cities that operate efficiently and purposefully. They are looking for smarter cities' (IBM, 2012).

There is no definite definition of a smart city, but the basic question is that what we want a smart city to be (Angelidou, 2014; Townsend, 2014). We may define smart cities as those where smart people live. There are two ways of defining smart people. One is intelligent and prosperous, but individualistic, consumerist and confined to himself, while the other is active in reaching out, contributing and transforming the life of the underclass. In a similar vein, the smart city is not an isolated enclave or a gated city but a city that connects with its people and transforms their lives. Thus, it would be pertinent to mention that while cities transform people, people make cities. We need to be clear about how we use the technology to shape our cities and its consequences for the poor and the people at the margins (Albino et al, 2015). Further, as India has been rapidly urbanising, the fortunes of people hugely depend upon how we build our cities. At the same time, intervention by the government could be helpful in envisioning and shaping our cities. The policies and programmes of the central government during the last decade should be seen in this light.

In India, the urban development programmes under JNNURM were redesignated into the Smart Cities Mission and AMRUT in 2015. The JNNURM was significantly different from earlier urban policies and programmes because it recognised the importance of cities in India's economic growth in the light of the emerging fact that a huge share of GDP accrued from urban areas (Bhagat, 2011). However, the urban development strategies like the Smart Cities Mission and AMRUT also reflect the continuity of the same ideology. Many criticised JNNURM for being exclusionary given a huge socio-economic inequality within the cities as well as between the regions. Inclusiveness has been a challenge for any urban development programmes along with growth and sustainability. The same criticism is also being levelled against smart cities' mission as well because of the similarity of their ideology promoting development based on privatisation, globalisation and neo-liberalism[1].

[1]Neoliberalism is a political economic theory and practices that believe that human well-being can best be advanced by liberating individual entrepreneurial freedoms and skills within an institutional framework in favour of strong private property rights, free markets and free trade. The role of the state is to create and preserve the institutional framework appropriate to such practices and keep its role minimum in the areas of welfare provisions and social security programmes (Harvey, 2005).

The Smart Cities Mission was initiated for 100 selected cities spread over across the country. The Mission aims to harness the potential of the city which aspires to become smart through smart solutions. Smart solutions include e-governance and electronic service delivery, video crime monitoring, smart metres for water supply management, smart parking and intelligent traffic management to mention a few from a long list. Application of smart solutions will enable cities to use technology, information and data to improve infrastructure and services. It will also endeavour area-based development through retrofitting (city improvement) and redevelopment (city renewal). In addition, new areas/greenfields (city extension) will be developed around the city to accommodate growing urban population. It is envisaged that the strategies for the development of a smart city will create enough jobs and take care of the poor. Thus, it is conceived that the smart cities would be inclusive.

The implementation of the Smart Cities Mission is carried out by a Special Purpose Vehicle (SPV) headed by a full-time CEO with nominees of Central, State and Local Governments. The SPV is a limited company under the Companies Act, 2013 at the city level. At the city level, a Smart City Advisory Forum is envisaged to advise and enable collaboration among various stakeholders and will include the District Collector, MP, MLA, Mayor, CEO of SPV, local youths and citizens and technical experts. The Smart Cities Mission requires smart people to participate actively in governance and reforms. The smart people are imagined as those increasing the use of ICT, especially mobile-based tools facilitated by SPV. The central government provides a very small grant initially with an equal matching grant from the state government. The future grants to the smart city depend upon performance. The 100 smart cities were selected based on competition inviting Smart City Proposal. A large number of consulting firms, as well as handholding agencies, are being engaged in different stages of smart city development. In the long run, these cities will acquire a brand and an identity based on their main economic activity such as local cuisine, health, education, arts and crafts, culture, sports goods, furniture, hosiery, textile, etc. Thus, the smart cities are envisaged to emerge not only as sites of production and efficient governance but also as sites of consumption. In the event of this, it is likely to spur economic growth and improve the quality of life of its citizens (Bhagat, 2015).

A complementary to the Smart Cities Mission, AMRUT is launched to cover 500 cities with a population of one hundred thousand and more. The mandate of the AMRUT confines to water supply, sewerage and septage management, storm water drainage, urban transport and development of green spaces and parks including capacity building and reform implementation by the Urban Local Bodies (ULBs). It is envisaged that the funding under AMRUT will give first preference to the potential smart cities. A State Annual Action Plan will be prepared (SAAP) in conformity with other central and state government programmes. The state contribution to SAAP should not be less than 20 per cent of the total project cost. The preparation of SAAP will be followed by the preparation of SLIP (Service Level Implementation Plans). Under AMRUT, one of the innovative approaches suggested for augmenting water supply is the recycling and reuse of water instead of bringing it from long distances. These two programmes also have a strong complementarity in achieving urban transformation. While AMRUT follows a project-based approach, the Smart Cities Mission follows an area-based strategy. However, both programmes endeavour to promote partnership with State, ULBs and private sectors with the central government playing a decisive role. However, it may be noted that both the Smart Cities Mission and AMRUT do not cover all cities and towns (about 8000 urban centres) of the country. Therefore, urban development as envisaged by the central government has been a selective process (see for details www.smartcities.gov.in; www.amrut.gov.in).

## 9.4 Reforms in Urban Governance and Need for Urban Planning

This amendment to the Constitution provided a common framework for the structure and mandate of urban local bodies to function as effective democratic units of local self-government and promote planning and urban development. One of the significant reforms that have been initiated by the central government is the promotion of decentralised local governance by the urban local bodies through the 74th Amendment to the Constitution effected in 1992. The amendment received the assent of the President of India on 20 April 1993. The amendment envisages planning and development of urban centres by urban local bodies which would be possible only through political, administrative and fiscal empowerment of them by the respective state governments. The amendment noted that 'in many States local bodies have become weak and ineffective on account of a variety of reasons, including the failure to hold regular elections, prolonged supersession and inadequate devolution of powers and functions. As a result, Urban Local Bodies are not able to perform effectively as vibrant democratic units of self-government'. The amendment also inserted the Twelfth Schedule in the Constitution mentioning the areas of functions of urban local bodies. The Twelfth Schedule mentions the following functions:

1. Urban planning including town planning.
2. Regulation of land use and construction of buildings.
3. Planning for economic and social development.
4. Roads and bridges.
5. Water supply for domestic, industrial and commercial purposes.
6. Public health, sanitation conservancy and solid waste management.
7. Fire services.
8. Urban forestry, protection of the environment and promotion of ecological aspects.
9. Safeguarding the interests of weaker sections of society, including the handicapped and mentally retarded.
10. Slum improvement and upgradation.
11. Urban poverty alleviation.

12. Provision of urban amenities and facilities such as parks, gardens and playgrounds.
13. Promotion of cultural, educational and aesthetic aspects.
14. Burials and burial grounds; cremations, cremation grounds and electric crematoriums.
15. Cattle pounds; prevention of cruelty to animals.
16. Vital statistics including registration of births and deaths.
17. Public amenities including street lighting, parking lots, bus stops and public conveniences.
18. Regulation of slaughterhouses and tanneries.

It was, however, noted by several experts that the implementation of 74th Amendment to the Constitution remains halfhearted as state governments have not fully empowered the urban local bodies to carry out the tasks as per constitutional provisions. On the other hand, various central government programmes were not much successful notably due to lack of single control, ownership and convergence of urban development programmes at the local level (HPEC, 2011: 23).

India's planning process has been centralised, top down and sector based. Same is true for urban development planning as well which focuses specifically on urban housing, water supply, sanitation, slum improvement, urban infrastructure, megacity, small and medium towns, etc. Although urban development is a state subject, there is little that the state could think about an integrated urban development strategy and planning. State Planning Commissions/Boards have hardly come out with state urban development strategies and are mostly dependent on what central government envisages through its mega urban development programmes. There is also a lack of convergence in various urban development programmes initiated by different ministries and government organisations. Earlier at the central level urban affairs are looked after by two ministries namely the Ministry of Urban Development and the Ministry of Housing and Urban Affairs. Since 2017 both Ministries were merged and now it is known Ministry of Housing and Urban Affairs (MoHUA). Pradhan Mantri Awas Yojana—Urban (PMAY-U) and Swachcha Bharat (Urban) are also looked after by this Ministry.

The potential of Smart Cities Mission and its convergence with AMRUT, Pradhan Mantri Awas Yojana—Urban (PMAY-U) and Swachcha Bharat (Urban) may bring many benefits, but there is a need to safeguard the poor and also the slum dwellers that comprise about one-fifth of the total urban population. These programmes should not be seen in isolation either at the level of governance or at the level of implementation; otherwise, they might lose sight of inclusiveness. As envisaged, smart cities are not meant to widen the digital divide but to help bridge the intra-urban gap as well as the rural–urban divide. The emergence of the idea of the smart city ought to be seen in the backdrop of the need to contain corruption and meet the requirement of adequate and efficient service delivery in urban areas using information and digital technologies. The success of the programmes will be judged in future by their power to transform the life of the people, and their ability to reduce growing inequality in our society.

Earlier committees have found that the funds allocated under various programmes are too low. For example, the High-Powered Expert Committee estimated an investment requirement of 2 trillion rupees per year over next the 20-year period (HPEC, 2011: 35). Further, programmes are also recycled by changing names due to political reasons creating confusion not only in its implementation but also in their impact assessment. For example, Jawahar Rozgar Yojana (JRY) was changed to Swaranjayanti Sahari Rozgar Yojana (SJSRY) and at present known as the National Urban Livelihood Mission (NULM). Similarly, RAY was renamed as PMAY-U.

On the other hand, urban development is a matter of utmost importance as urban areas are constantly stressed for civic amenities like access to electricity, drinking water, sanitation, LPG, etc. Road congestion, traffic, air and water pollution, municipal waste disposal and law and order are other issues of concern for urban governance. Urban planning has to be done by the urban local bodies which comprise mainly Municipal Corporations, Municipalities and Nagar Panchayats controlled by the state government. Urban local bodies need to be empowered with a clear delegation of functions, financial resources and autonomy. There is also a need to enhance the administrative, managerial and technical capacity of the urban local bodies.

According to NITI Aayog (2021) about 52 per cent of the statutory towns and 76 per cent of the census towns do not have any Master Plans to guide their spatial growth and infrastructural investments. This implies that three-fourths of the urban centres in the country do not have any spatial strategy for the next 20 years. It must be clearly recognised that urban planning cannot be divorced from urban governance. For large urban agglomerations, which are spread over several municipal bodies and interspersed with villages, planning should be taken care of by the Metropolitan Planning Committee to be constituted as per the provision of 74th Amendment to the Constitution.

There is a serious lack of effective urban planning in India for most of the cities. For many big cities for which plans have been prepared are outdated and do not reflect the concerns of the urban dwellers in general and urban poor in particular. Even where plans are updated and newly prepared, they are prepared by consultants and technical experts, sometimes even by outsourcing. Planning in a real sense should involve the urban dwellers to accommodate their needs and concerns; it should be participatory and decentralised. Why it is not so? The answer lies in the nature of urban governance. As urban bodies lack power and resources, parastatal bodies like the Urban Development Authority have virtually taken over the city promoting real estate and infrastructure. On the other hand, basic services like water, sanitation, education, health care, shelter, etc., are neglected. As urban development is a state subject, the state governments must realise the role of the city and initiate urban reforms to empower local bodies both administratively and fiscally. We must envision cities as engines of human development rather than capital accumulation by the few.

There is a need to democratise and empower urban local bodies as per the provisions of 74th Amendment to the Constitution. Mayor should control the city both in power and functions. The institution of Mayor/Mayor in Council should be made responsible for the city. Many global cities have addressed the problems of governing large cities through the elected local government headed by Mayor. The Mayor of

London and New York, for example, are directly accountable for strategic governance of their respective cities and have substantial resources and political power at their disposal. The municipal councillors must play a role in ensuring development in their respective wards, although they are not necessarily best placed to take important strategic decisions for the city. Devolution of powers to the Mayor could be a significant first step. Besides, this also makes the Mayor as the primary stakeholder. Facing a city's diverse electorate at the end of every term can go a long way in ensuring most development aspects are looked at and managed effectively. The key issue is that local democracy may be revitalising if it is also accompanied by financial empowerment and resources. Democratisation and empowerment of urban local bodies will pave the way for new planning culture which is need based, participatory and decentralised with an aim to improve the quality of urban life and promote sustainable urban development.

Urban development is a state subject in India and merely initiation of urban reforms and urban development programmes are not sufficient unless implemented by the state governments and urban local bodies. Whereas central government provides guidelines and also promises increased funds through centrally initiated urban development programmes, but this needs to be implemented by the state governments. Similarly, the implementation of JNNURM has been tardy and poor. On the other hand, while many small towns are still governed by rural local bodies (*panchayats*) (Bhagat, 2005), there are multiple agencies responsible for the planning, development and governance in the big cities. Sometimes one wonders who owns the city? For example, in Mumbai, there are a host of parastatal bodies like the Mumbai Metropolitan Region Development Authority (MMRDA), Maharashtra Housing and Area Development Authority (MHADA), Slum Rehabilitation Authority (SRA), City and Industrial Development Corporation (CIDCO) which are looking after various activities apart from Municipal Corporation of Greater Mumbai (MCGM). Further, Mayor and elected councillors are not the decisive in civic administration vis-à-vis the role of Municipal Commissioner. Further in most cases, the state governments have not yet constituted the Metropolitan Planning Committee as envisaged in the 74th Amendment to look after the planning and development beyond the jurisdiction of respective local bodies to develop and govern the whole areas under the metropolitan region. As such, there is a lack of local democracy and empowerment of urban local bodies both politically and fiscally. Though local governance could have some problems initially such as inefficiency and corruption as pointed out in some studies, yet in the long run there seems to be no alternative to the process of democratisation for planning and development of urban areas and solution to the urban problems which are place specific (HPEC, 2011). The twin process of democratisation and empowerment of municipal governance not only meets the challenges of speedier urbanisation like shortage of civic amenities, affordable housing, health care needs and poverty alleviation but also promotes the political and economic inclusion of marginal communities like poor, migrants and slum dwellers within the urban centres.

## 9.5 Urbanisation Policies for Rural Development:

Urban policies in India highly focused on urban development policies rather than an urbanisation policy for rural development. The true strength of urbanisation lies in how it could become a vehicle for rural development. Historically rapid rural to urban transition accompanied the economic development in Western Europe (Lewis, 1954; Rostow, 1960). Rural to urban migration was rapid which led to massive transfer of surplus labour from rural to urban areas. Many Western European countries also experienced massive emigration to their respective colonies. In Britain nearly half the increment in the population every year was migrating for permanent settlement abroad during 1821 to 1915. Britain alone accounted for 37 per cent of all emigration from Europe (Patnaik, n.d.; Kenwood & Lougheed, 1971). Such an opportunity of exporting a substantial proportion of unemployed persons abroad never existed for most of the developing countries. On the other hand, in many developing countries including India, the contribution of migration in urban transition is hugely low—i.e. only one-fifth of urban population growth is attributed to net rural to urban migration, while villages acquiring in situ urban characteristics has been a growing phenomenon popularly known as emergence of census towns reflecting rural to urban transition in the form of in situ urbanisation. This indicates that country like India should also focus attention on urbanisation policies along with urban development policies. There is a need to make a distinction between urbanisation policies and urban development policies to understand the process of urbanisation unfolding at a larger geographical scale and not only confined to the precincts of big cities. On the other hand, urban development policy aims at raising the income level, employment opportunities and access to civic services in the existing agglomerations urban centres in order to improve urban quality of life. In the context of India, an urban policy perspective hardly makes distinction between the two because urbanisation is viewed as synonymous with urban development which is biased towards big cities. It is worthwhile to emphasise that urbanisation policy means how to transform rural into urban areas through the provision of infrastructure, while urban development policy means improving human conditions within the city/urban areas. Improving human conditions and productivity within the city is potentially linked to developing rural conditions and output as well. It was pointed out by HPEC (2011: 2) that funding of rural development programmes would simply not happen unless cities develop and generate revenues of central, state, and local governments for both urban and rural development. This is the basic argument for urbanisation as an engine of economic development in general and rural development in particular. However, urban development alone does not constitute urban policy as urbanisation policy is equally important for the dispersal of the process of development in achieving balanced regional development.

It is very important to emphasise that rural to urban migration plays a very important role in rural to urban transition. However, a significant proportion of rural to urban migration in many developing countries including India has been temporary and circular in nature with strong rural ties, and migrants be they long-term or short-term keeping shuttling between rural and urban areas (Bhagat & Keshri, 2024). Migrants mostly work in the informal sectors with low wages and extremely insecure life conditions, and rural to urban transition is not similar to what is historically experienced in the west European countries. As a result, India's economic growth trajectory is laden with a huge number of her workforce dependent on agriculture located in rural areas. According to NITI Aayog, 45 per cent of India's workforce is engaged in agricultural and allied activities contributing to just 18.6 per cent of the national income at current prices during 2021–2022 (NITI Aayog, 2023).

Lack of desired urban transition coupled with increasing opportunities for demographic dividend due to huge fertility decline in India warrants that the urban transition ought to focus on inclusive and sustainable urbanisation policies that are not merely confined to increasing GDP, but truly act as a vehicle of rural development. This requires that the urban policies of India should focus on the inter-linkages between urban and rural areas, and must try to bridge the urban–rural divide. The development of small and medium towns and a strategy of PURA can play a very important role.

There has been an ambivalent approach to develop the small and medium towns, and the funding has been extremely low. When AMRUT started in 2015, the existing projects under UIDSSMT were funded until 2017 and later handed over to the states/UTs (PIB, 2023). Many state governments do not have capacities, resources and vision to undertake development of small and medium towns and leverage them as engines of rural development. A partnership between the central and state governments with a vision of sustainable development accompanied by a spatial strategy focusing on small and medium towns would be certainly important in achieving not only rural development but also balanced regional development within the country.

Perhaps the only policy that can be truly called an urbanisation policy is PURA (Providing Urban Amenities to the Rural Areas). The idea of PURA was mooted by the former President of India in the address to the Nation on the eve of Republic in 2003.[1] As a sequel, the Ministry of Rural Development initiated the PURA scheme on a pilot basis in several parts of the country in 2004–2005.

The scheme continued in various forms subsequently but remained in pilot mode only. A modified version of PURA was introduced in 2012 based on public–private partnership envisioning a critical role of a private partner who shall be selected on the basis of bidding responsible for developing the PURA project in a panchayat and a cluster of panchayats. The idea of PURA further evolved into a Rurban Mission known as Shyama Prasad Mukherji Rurban Mission launched in 2016. National Rurban Mission is an organised development of the clusters of villages, the areas would be well delineated and would be duly notified as planning areas under the relevant Planning Acts of the States.

[1] The President of India, Dr A.P.J. Abdul Kalam, Developed India: Vision and Actions, Address to the Nation on the eve of Republic Day 2003. https://archive.pib.gov.in/archive/releases98/lyr2003/rjan2003/25012003/r250120034.html. Accessed on 2 March 2024.

It is important to reiterate that in spite of continued effort, the implementation of PURA as envisioned by our former president has not met with desired success. PURA essentially is an instrument of rural transformation leveraging urbanisation as a vehicle of rural development. It is based on the criteria of connectivity, capacity and co-operation in rural setting. It consists of four components namely physical connectivity by providing roads in rural areas, electronic connectivity by providing a reliable communication network and knowledge connectivity by establishing more professional institutions and vocational training centres. All this connectivity needs to be done in an integrated way so that economic connectivity will emerge leading to self-actuating people and economy. As opportunities and challenges vary from one part of the country to another, it is suggested that depending upon geographical conditions PURA can be designed as Plain, coastal, desert, hill PURA, etc. It can consist of 20 to 45 villages with a population as high as 100 thousand. It is also suggested that India can have 7000 PURAs covering more than 6 hundred thousand villages (Kalam & Singh, 2011).

The idea of PURA could be combined with the growth pole/growth centre approach. A growth pole/growth centre could be a census town or a small and medium town as a core surrounded by a cluster of villages. This will help improving connectivity, generate spatial efficiency and create agglomeration economies much needed for attracting investment and promoting entrepreneurial activities for rural transformation.

**Acknowledgements** This chapter draws upon two earlier articles published by the first author, namely, 'Urban Policies and Programmes in India', YOJANA, September 2014, pp. 4–8, and 'Smart City: Concept and Strategies', YOJANA, September 2015, pp. 7–9. This is a completely revised version incorporating recent changes in the urban policies and programmes in India.

## References

Albino, V., Berardi, U., & Dangelico, R. M. (2015). Smart Cities: Definitions, Dimensions, Performance, and Initiatives. *Journal of Urban Technology, 22*(1), 3–21. https://doi.org/10.1080/10630732.2014.942092

Angelidou, M. (2014). Smart City Policies: A Spatial Approach. *Cities, 41*, S3–S11. https://doi.org/10.1016/j.cities.2014.06.007

Anthopoulos, L. G., & Vakali, A. (2012). *Urban Planning and Smart Cities: Interrelations and Reciprocities*. http://www.researchgate.net/publication/230851445_Urban_Planning_and_Smart_Cities_Interrelations_and_Reciprocities

Bhagat, R. B. (2005). Rural-urban classification and municipal governance in India, *Singapore Journal of Tropical Geography*, *26*(1), 61–73.

Bhagat, R. B., & Keshri, K. (2024). Internal Migration and Temporary Labour Circulation in India. In K. S. James & T. V. Sekher (Eds.), *India Population Report* (pp. 246–269). Cambridge University Press.

Bhagat, R. B. (2014, September). Urban Policies and Programmes in India, pp. 4–8. YOJANA.

Bhagat, R. B. (2015, September). Smart City: Concept and Strategies, pp. 7–9. YOJANA.

Bhagat, R. B. (2011). Emerging Pattern of Urbanisation in India. *Economic and Political Weekly, 46*(34), 10–12.

Bhagat, R. B., & Mohanty, S. (2009). Emerging Pattern of Urbanization and the Contribution of Migration in Urban Growth in India. *Asian Population Studies, 5*(1), 5–20.
Bose, A. (1980). *India's Urbanization 1901–2001*, 2nd revised ed. Tata McGraw Hill Publishing Company Limited.
Harvey, D. (2005). *A Brief History of Neoliberalism.* Oxford University Press.
HPEC (High Powered Expert Committee). (2011). *Report on Indian Urban Infrastructure and Services.* Ministry of Urban Development, Government of India.
IBM. (2012). How to Transform a City: Lessons from the IBM Smarter Cities Challenge. IBM Smarter Cities White Paper. http://asmarterplanet.com/files/2012/11/Smarter-CitiesWhitePaper_031412b.pdf
Kalam, A. P.J. Abdul, & Singh, S. P. (2011). *Target 3 Billion, PURA: Innovation Solutions Towards Sustainable Development.* Penguin Books.
Kenwood, A. G., & Loughheed, L. (1971). *The Growth of the International Economy 1820–1960.* Allen and Unwin.
.
Lewis, W. A. (1954). Economic Development with Unlimited Supplies of Labour. *The Manchester School, 22*(2), 139–191.
Niti Aayog. (2021). Reforms in Urban Planning Capacity in India: A Final Report, National Institute for Transforming India (Niti Aayog), Government of India, New Delhi.
Niti Aayog. (2023). From Green Revolution to Amrit Kaal: Lessons and Way Forward for Indian Agriculture, Niti Aayog Working Paper 02/2023, National Institute for Transforming India (Niti Aayog), Government of India, New Delhi.
Patnaik, U. (n.d). *Capitalist Trajectories of Global Interdependence and Welfare Outcomes: The Lessions of History for the Present.* https://www.networkideas.org/ideasact/dec11/pdf/Utsa%20_Patnaik.pdf, accessed on 1st Match 2024.
PIB (Press Information Bureau). (2023). Urban Infrastructure Development Scheme, Government of India. https://www.pib.gov.in/PressReleseDetailm.aspx?PRID=1910166, accessed on 1st March 2024.
Planning Commission, Government of India. (2008). *Eleventh Five Year Plan (2007–2012), Vol III: Agriculture, Rural Development, Industry, Services and Physical Infrastructure* (pp. 394–422). Oxford University Press.
Planning Commission, Goverment of India. (2010). Reference Material 2010, Notes on the Functioning of Various Divisions, Planning Commission, Government of India.
Planning Commission, Government of India. (2013). *Twelfth Five Year Plan (2012–2017), Vol II: Economic Sectors* (pp. 318–361). Sage Publication India Pvt Ltd.
Ramachandran, R. (1989). *Urbanisation and Urban Systems in India.* Oxford University Press.
Rostow, W. W. (1960). *The Stages of Economic Growth: A Non-Communist Manifesto.* Cambridge University Press.
Sanyal, A. (2018). The Making of a Mythical Forerunner. In S. Baru & M. Desai (Eds.), *The Bombay Plan: Blueprint for Economic Resurgence.* Rupa Publications India Pvt. Ltd.
Shelton, T., Zook, M., & Wiig, A. (2015). The Actually Existing Smart Cities. *Cambridge Journal of Regions, Economy and Society, 8*, 13–25.
Shaw, A. (1996, January 27). Urban Policy in Post-independent India: An Appraisal. *Economic and Political Weekly* (pp. 224–228).
Times of India. (2008, June 13). *National Daily* (p. 16).
Townsend, A. (2014). *Smart Cities.* W.W. Norton & Company Inc.

# Epilogue

Ontologically, urbanisation means how spaces and places are transformed by the activities of people. Urbanisation is about increasing the concentration of population and economic activities in settlements characterised by the manifestation of agglomeration economies. Historically, it resulted from the transformation of an agrarian economy into a non-agrarian one. Diversification of the economy, division of work, increased specialisation of work and skills, advancement of knowledge, improvement in transport and communication, changes in the nature of governance and increasing role of science and technology are concomitant changes associated with it, and also shaping urbanisation. It appears that industrialisation has been a precondition of urbanisation. However, this is not true in the contemporary time. In fact, urbanisation has emerged as a complex socio-economic and political process that defies urban definition and the rural and urban categorisation (Brenner & Schmid, 2015). Notwithstanding, countries across the world consider urbanisation as a characteristic of human settlement and measure it in terms of size, density, occupation, status of governance and spatial characteristics connoting various types of infrastructure and amenities. This book reflects this empirical reality of urbanisation, its close relationship with economic development and challenges associated with inclusive urbanisation and sustainable development.

Urbanisation is one of the mega-trends along with population ageing, rising inequalities, climate change and predominance of technology such as AI (artificial intelligence). All these mega-trends are interrelated and reflected in our lives in an increasingly urban world. More than half of the world's population now lives in urban areas (United Nations, 2019). Also, within urban areas, there is a growing concentration of population in metropolitan and megacities. This profound change at the global level has manifested more rapidly in less developed countries. This has raised many questions about the nature, patterns, trends and characteristics of urbanisation having important bearings on the well-being of people and their quality of life; growing inequality between rural and urban areas and between regions; issues



R. B. Bhagat and M. I. Hassan, *Urbanisation and Urban Policies in India*, Exploring Urban Change in South Asia, https://doi.org/10.1007/978-981-96-3181-0

pertaining to access to resources, energy consumption and environmental degradation; and last but not the least the problems of local governance. Dynamics in urban population along with trends and characteristics of migration, in general, and rural to urban migration, in particular, are important correlates of unfolding economic and social changes, which are basically urban in character.

India is now the most populous country in the world and is urbanising rapidly. Our understanding of the process of urbanisation that was earlier negatively loaded has given way to a positive realisation in the policy and planning spheres recently. This change in thinking has come about because urban areas by the early 2020s contributed about 60% of the GDP, and there is a widespread realisation that the ambitious growth in GDP cannot be achieved without a vibrant urban sector (Niti Aayog & Asian Development Bank, 2022). At the same time, urbanisation has been considered as one of the major challenges, requiring a massive expansion in urban infrastructure and services (Ahluwalia et al., 2014). Therefore, an understanding of the emerging patterns of urbanisation, the growth of urban centres by size class, the linkages between urbanisation and migration, the nature and patterns of migration and the issues related to governance, urban policy and programmes assume enormous significance. This book addresses these and similar other concerns in its nine chapters. The book heavily relies on available official data and emerging literature on urbanisation in India in order to understand the strengths and weaknesses of the urbanisation process from a spatial perspective and critically evaluates urban policy and programmes in the light of its emerging features and the goals of inclusive and sustainable urbanisation.

Chapter 1 highlights that urbanisation has been a historical as well as a spatial process. It discusses the nature and levels of urbanisation right from the ancient times to the present as a process of spatial transformation linked to development and social organisation. It also emphasises the nature of colonial urbanisation that coincides with industrial development and expansion of service sectors in India and several other developing countries. For this period there is a plenty of data available on urbanisation from the censuses of various countries. As such this chapter highlights the differences in the nature of urbanisation in the developed and less developed parts of the world and attempts to contextualise India's urbanisation from a global perspective.

The definition of 'urban' and scheme of rural–urban classification as in vogue in different countries including India are examined in Chapter 2. This chapter shows that the criteria of urban differ from country to country and there is no universal definition of 'urban' at the global level, although some recent attempts have been made for an identical definition. Diversity of criteria is a very important aspect of definition of urban which reflects the global diversity in the social and economic conditions. Nevertheless, most of the countries have evolved their own definition of urban and rural–urban classification constitutes an important framework for collecting and compiling population data. While 'urban' is often specifically defined, 'rural' is treated simply as the residual category. This chapter also demonstrates that rural and urban classification is highly significant for local governance and development planning. In the Indian context, the constitutional amendments have given local bodies a legitimate role in the federal structure of the country. While rural–urban classification is the task of the Census of India, state governments are responsible for granting municipal status to urban centres. While a large number of census towns, or non-municipal towns, are identified by the Census of India, the respective state governments have not granted municipal status to many of them. It appears that while municipalisation has been an integral component of urbanisation, urbanisation may not lead to municipalisation at the earliest.

Chapter 3 presents levels, trends and patterns of urbanisation in India since the beginning of the twentieth century. The chapter focuses on the measurement of the processes of urbanisation based on demographic data. Measures like the degree, tempo and scale of urbanisation are employed to study the level, trend and pattern of urbanisation. It also tries to estimate the contribution of different components, namely, natural increase, net rural to urban migration and net rural to urban classification in urban population growth to reflect the relationship between economic development and urbanisation. The chapter finally examines the level of regional economic disparities in India and its association with urbanisation.

As an epithet of urbanisation, the urban size is the most important attribute influencing various demographic and socio-economic conditions of an area. The concentration of population in big cities has been associated with an increasing urban agglomeration economy. It is a matter of inquiry as to how bigger cities are growing compared to intermediate and small urban centres. The growth dynamics of urban centres by size class are closely related to the urbanisation cycle. In the early phase, bigger cities are likely to have a higher growth rate, but in the latter phase, small towns are expected to grow as a result of increased congestion and crowding in the large and intermediate towns. Chapter 4 throws light on the role of urban population growth rate by size class of urban centres in urban transition. This chapter further critically examines the rank-size rule as a theoretical framework for urban efficiency. As the size class of urban centres is accumulative and promotes concentration, it ignores the issues related to equity, inclusion and inequality as linked to the rank-size rule in the existing literature. This chapter tries to address 'efficiency vs. equity' concerns in the urban system influenced by the contemporary processes of urbanisation.

Chapter 5 presents the contribution of migration as a significant component of urbanisation, particularly the one from rural to urban areas. Although increasing over time, urban to urban migration is inconsequential from the point of view of urban growth. This chapter highlights trends, regional patterns and characteristics of rural to urban migration in relation to development. Reasons for migration by gender form an important part of this chapter.

Urbanisation in India is characterised by metropolitisation and this issue has been taken up in Chapter 6. Out of 52 million plus cities as per the 2011 census, 8 of them could be defined as megacities with a population of 5 million and more. As these cities dominate the urban and economic space, this chapter focuses on the population growth and characteristics of these cities. It is also a known fact that the metropolitisation of India's urban space is accompanied by urban sprawl and peri-urbanisation of the adjoining areas around the megacities. A core-periphery framework is used in this chapter to study the pattern of migration. It may be recalled that during the COVID-19 times, these megacities impelled a large number of migrant workers revealing urban inequality and urban exclusion as a characteristic feature of India's urbanisation.

Urbanisation is closely associated with improved income levels and living standards of people across the world including India. Urbanisation is, thus, expected to improve quality of life in terms of housing conditions and access to basic amenities like electricity, clean fuel such as LPG/PNG, safe drinking water, improved sanitation, etc. Chapter 7 focuses on the above-mentioned basic amenities and tries to examine the variations at the state level and by size class of urban centres and its administrative status.

The process of urbanisation manifests itself in various forms of inequality, for instance, in income levels, access to health care, education and labour market and slum and housing conditions. The economic and social inequality is also visible in the social spaces of the city in the form of residential segregation. Enclaves of communities, slums and gated communities are some of the prominent forms of segregation of social space in the cities. At the same time, private cities are also emerging as spaces of consumption and repositing wealth. Inequality and segregation in this situation entail tension and conflicts both in latent and manifested forms. As a result, various forms of crime and violence become an integral part of urban life. Chapter 8 examines inequality and segregation as an emerging feature of exclusionary urbanisation and its linkages with crime and violence. The chapter indicates that the securitisation of the city using technology like sensors, cameras, drones and GPS has entered the production and planning of the urban spaces of cities ignoring human and social issues. The increasing role of digital technology and AI (artificial intelligence) could either ameliorate or mar the life of people if digital and technological divides worsen in future.

Sustainable and inclusive urbanisation is a critical factor in order to achieve SDG goals. Goal 11 is specifically concerned with Sustainable Cities and Communities. In this respect, urban policies and programmes are very important to achieve this goal. In Chapter 9, an attempt has been made to distinguish between urban development policies, which are focused on cities and their problems, and urbanisation policy which addresses the economic, social and environmental linkages between urban and rural areas by strengthening national and regional development planning. The purpose of urbanisation policy is to reduce the rural and urban divide and act as a vehicle for rural development. On the other hand, urban development policies and programmes have a distinct focus on towns and cities. Although there are some programmes launched under the *National Rurban Mission* as part of the urbanisation policy, its architecture and funding come from the Ministry of Rural Development and have very limited impact in the countryside.

In a nutshell, the book attempts to discuss the various facets of urbanisation unfolding over space and time primarily focusing on demographic changes in India. Keeping in view the ontological premise of defining urbanisation as changing relationship between people and place is an inevitable process. Further, it is also worthwhile to note urbanisation is not a finite process as we believe in the conventional measurement of urbanisation based on a percentage of urban. Moreover, urbanisation is about how we organise our human settlements for work, living, mobility and sustainability. It is about human capability rooted in inclusive agglomeration enhancing efficiency, productivity and equity. It needs to be harnessed and must not be left entirely to the market forces which may not ensure equity while unleashing the forces of efficiency and productivity. It could also take a cyclical form of urbanisation, followed by deurbanisation and reurbanisation. In this sense, public policy assumes an enormous role in shaping inclusive and sustainable urbanisation, mitigation of climate crisis and its harmonisation with nature.

In the words of former U.N. Secretary-General Ban Ki-moon, 'Our struggle for global sustainability will be won or lost in cities' (United Nations, 2012). In this context, 'Right to the City' could provide a philosophical and theoretical basis for people-centric urbanisation towards achieving the goals of sustainable development and meeting the challenges arising from demographic shift, climate change and artificial intelligence.

## References

Ahluwalia, I. J., Kanbur, R., & Mohanty, P. K. (2014). *Urbanisation in India: Challenges, Opportunities and the Way Forward*. Sage Publication Pvt.

Brenner, N., & Schmid, C. (2015). Towards a New Epistemology of Urban? *City, 19*(2&3), 151–182.

Niti Aayog and Asian Development Bank. (2022). *Cities as Engines of Growth: Strengthening the States for Broad Based Urban Development*, https://www.nitiforstates.gov.in/policy-viewer?id=RNTNAM000397, accessed on August 19, 2024.

United Nations, Department of Economic and Social Affairs, Population Division. (2019). *World Urbanization Prospects 2018*: Highlights (ST/ESA/SER.A/421).

United Nations. (2012, April 23). Meeting Coverage and Press Releases, Secretary-General SG/SM/14249 ENV/DEV/1276 HAB/217, https://press.un.org/en/2012/sgsm14249.doc.htm, accessed on September 28, 2024.

MIX
Papier aus verantwortungsvollen Quellen
Paper from responsible sources
FSC® C105338

If you have any concerns about our products,
you can contact us on
**ProductSafety@springernature.com**

In case Publisher is established outside the EU,
the EU authorized representative is:
**Springer Nature Customer Service Center GmbH**
**Europaplatz 3, 69115 Heidelberg, Germany**

Printed by Libri Plureos GmbH
in Hamburg, Germany